Firefighters and Fires in Providence

A Pictorial History of the Providence Fire Department
1754-2001

A little fire is quickly
    trodden out;

Which, being suffer'd, rivers
    cannot quench.

King Henry VI, Part 3, Act 4, Scene 8
William Shakespeare

Firefighters and Fires in Providence

A Pictorial History of the Providence Fire Department 1754-2001

by Patrick T. Conley and Paul R. Campbell

Design by
Paula Hennigan Phillips

Published jointly
by
The Rhode Island Publications Society
and
The Donning Company/Publishers
Virginia Beach

Copyright © Rhode Island Publications Society 1985, 2002
All rights reserved
Printed in the U.S.A.
Library of Congress Catalog Card Number 85-62048
ISBN: 0-917012-79-8
Second Edition 2002

Typeset, designed and produced by
The Donning Company/Publishers
184 Business Park Drive, Suite 206
Virginia Beach, VA 23462-6533

To those Providence firefighters who have made the ultimate sacrifice in the performance of their duty.

Joshua Weaver (March 1828)
John H. McLane (September 1870)
Alexander J. McDonald (October 1883)
Daniel W. Brown (October 1883)
Nicholas C. Duff (December 1886)
Captain Hiram O. Butts (January 1901)
Joseph Devine (July 1902)
John Carlin (July 1903)
Captain George H. Noon (January 1907)
Benjamin N. Brown (December 1908)
Lieutenant Christopher Carpenter (May 1912)
Harry H. Howe (May 1912)
Captain Michael J. Kiernan (February 1921)
Edward H. Kelleher (February 1921)
John Teague (February 1921)
Arthur Rooney (February 1921)
Charles Keegan (June 1930)
Richard Coleman (November 1930)
Robert F. McDonald (March 1932)
Julian Miluck (April 1936)
Raymond Winters (May 1938)
Captain William L. Smith (June 1943)
Lieutenant William J. McElroy (March 1944)
Raymond Dean (December 1944)
Chester King (July 1945)
Louis Defresne (April 1955)
Captain George O. Heustis (March 1958)
Norman Clark (April 1960)
George Magnan (June 1961)
Lieutenant Joseph F. Dorsey (March 1963)
Captain Francis A. Shea (May 1964)
Edouard R. Theriault (June 1964)
Alfred Millard (November 1965)
Battalion Chief Joseph J. Mainey (April 1967)
Earl T. King (June 1967)
Joseph H. LeBlanc (March 1970)
William J. Moreland (December 1977)

Contents

Appendices

This book is dedicated to
the 343 New York firefighters
who lost their lives in the
World Trade Center tragedy.

Introduction

In 1850, while the Providence Fire Department was still in its volunteer stage, a Providence carpenter and a local patternmaker formed a partnership which was later incorporated under the name of the Providence Steam and Gas Pipe Company. The original site of this fledgling business was a 2,400-square foot section of the new Providence and Worcester Railroad building at Providence, Rhode Island. The company's early products included water pipes and devices for heating with exhaust steam from Corliss engines, as well as plants for making gas from rosin, crude oil, and coal.

When mechanical engineer and Corliss Steam Engine Company superintendent Frederick Grinnell bought an interest in the company in 1869, the Providence Steam and Gas Pipe Company began its rise from local to international prominence. In 1878 Grinnell entered into an agreement with Henry S. Parmelee to manufacture Parmelee sprinklers on a royalty basis. Grinnell improved upon Parmelee's inventions, and in 1881 he invented and secured patents on a heat-sensitive automatic sprinkler that, according to one source, "completely revolutionized the system of fire protection in manufacturing establishments throughout the world." The new product prompted the creation of a firm called the General Fire Extinguisher Company. Grinnell constantly improved and refined his invention, earning it the respect and endorsement of all major fire insurance companies. An 1895 meeting between Grinnell and fire representatives of the insurance industry led to the creation of the National Fire Protection Association the following year.

Eventually this firm assumed the name of its most distinguished leader and became the Grinnell Corporation, a major Providence employer with manufacturing plants also in Cranston, Rhode Island; Oshawa, Ontario; Columbia, Pennsylvania; Warren, Ohio; and Atlanta, Geor-

gia; plus two dozen strategically sited warehouses throughout the United States and Canada as well. In 1971 Grinnell was acquired by the International Telephone and Telegraph Company. Two years later ITT was ordered to divest itself of Grinnell's fire protection division because such ownership conflicted with ITT's control of the Hartford Fire Insurance Company. Since January 1976, when Tyco Laboratories, Inc., a New Hampshire-based holding company, acquired Grinnell Fire Protection Systems Company, Inc., both companies have prospered and complemented each other.

Grinnell Fire Protection Systems now employs over three thousand people, some of whom are in Rhode Island at the firm's Providence headquarters and at a combined district office/research division in Cranston.

In 1981 Grinnell registered its latest industrial achievement—the first residential sprinkler to pass all Underwriters Laboratory tests.

Grinnell's long and continuing relationship with the Providence Fire Department has been cordial and mutually productive. The Providence Fire Department is the winner of four Grand Awards. Grinnell, too, is a recognized leader in fire prevention. For more than 130 years, Grinnell and the fire department have worked together for the citizens of our city and have guarded their safety. Providence's long-standing 2A fire insurance rating is another product of their joint effort.

It is with pride and pleasure, therefore, that Grinnell joins with the Providence Fire Department in producing and publishing this book on firefighters, fires, and fire prevention in our hometown.

We consider this work our tribute to the firefighters of Providence, past and present, and to the memory of our company's most notable contribution to the cause of fire prevention—Frederick Grinnell.

Foreword

In 1982, while addressing a banquet for retired Providence firefighters, I looked out into the audience and saw generations of firefighters who had served the people of this city for decades upon decades with loyalty, professionalism, and responsibility. I approached the speaker's podium and announced that day that the City of Providence would undertake to produce a history of its fire department as a tribute both to the men who have served as our firefighting force and to their families.

The next morning I contacted my Office of Community Development and asked that money be set aside so that a pictorial history of our fire department could be written. Realizing that this would be no easy project, I was pleased when MOCD selected Providence authors Dr. Patrick T. Conley and Paul R. Campbell for the job. This fine book is the result of their collaboration.

Firefighters and Fires in Providence traces the history of our department with great accuracy and with an understanding of the traditions and heritage of the second oldest professional fire department in America. The selection of photographs, the descriptive captions, and the crisp historical reporting make this book not only a pleasure to read but a keepsake for all of our citizens to possess.

In my years as mayor, I was especially proud of our fire force because of its professionalism and a certain camaraderie that exists in that department, a spirit of good-fellowship that is hard to find anywhere else in municipal service. The cordial and satisfying relationship that I enjoyed with the Providence Fire Department was a high point of my tenure as the city's chief executive.

I know that all firefighters will treasure this book, which recounts a part of their lives; that they will peruse it often; and that it will reinforce their great pride in a department that is second in quality to none.

Vincent A. Cianci, Jr.

Acknowledgements

This book, like any substantial undertaking, is the product of many hands. We had our volunteers, our call men, and our permanent paid force, but all who participated exhibited one common denominator—an enthusiasm for researching and writing about one of the nation's most historic and distinguished departments, the second oldest permanent paid force in the United States and a perennial leader in the area of fire prevention.

Designed to increase the sense of pride and *esprit de corps* that has always marked the Providence Fire Department and to generate awareness in the local community of our fire service's proud heritage, this project was suggested by Mayor Vincent A. Cianci, Jr., and carried forward by Chief Michael F. Moise, son and brother of Providence firefighters. Cianci recommended use of his option funds from the Community Development program to finance the effort.

Our apologies are extended to the privates, the carpenters, the mechanics, the repairmen, the supply room personnel, the salvage workers, and the clerks for slighting them in this study. As Chief Moise observed in his 1979 letter to the Home Rule Charter Commission, "The fire department is a quasi-military organization." And just as military histories concentrate on generals and field commanders, so also do we focus on chiefs and those officers engaged directly in firefighting or rescue work. In the main, they have set the policy, left the written record, and generated the media coverage upon which this historical narrative is based. But we in no way mean to disparage or minimize the toil, the dedication, or the indispensability of the department's rank and file—both present and past.

The authors, selected presumably for the expertise gained in writing a general history of Providence, found their task of researching 247 years of department history a challenge, yet a delight. Our principal auxiliaries were Patrick T. Conley, Jr., a Providence College history major, and Gail C. Cahalan, who became so absorbed in this project that she went out and purchased an old Providence fire station of her own. Their efforts in gathering a mountain of newspaper clippings dating back to 1801 and their role in helping us prepare the roster of notable fires are gratefully acknowledged.

Our picture research was aided by Maureen Taylor and Joyce Botelho, graphics curators at the Rhode Island Historical Society, an agency that gave us the run of its collections. Jeanne Richardson of the Providence Public Library and the Providence Heritage Commission assisted us with photos and clippings from the PPL's Rhode Island file, while Denise Bastien and Lou Notarianni were always available to photograph our glossy reproductions and current scenes. Mildred Longo of Warwick gave us the use of rare nineteenth-century documents from her personal collection, and Thomas Aldrich, descendant and namesake of the second chief of the paid department, assisted us with research on the life and times of his illustrious ancestor. Chief Moise gave us invaluable and detailed insights on the modern era, and William Carr patiently and helpfully escorted us to the various depositories containing the department's scattered records—headquarters, the training academy, and Chad Brown, where John McDonald of the Providence Permanent Firemen's Relief Association was a wealth of information.

The authors, either jointly or singly, visited several fire museums to obtain data, photos, and "hands-on" knowledge of the apparatus and equipment herein described, specifically the museums in New Bedford, South Carver (Edaville), and Brewster, Massachusetts; Chambersburg, Pennsylvania; and especially Hudson, New York, where the magnificent and spacious New York Fire Museum features, among its many impressive displays, an 1863 Pawtucket-made Jeffers engine.

Jan Campbell and Phyllis Cardullo performed the bulk of the typing chores with occasional and able assistance from Maria Lopes and Patricia Rekrut. Dr. Hilliard Beller of the Rhode Island Publications Society handled the copyediting and proofreading chores with his usual meticulous care. Finally, Grinnell Fire Extinguisher Company of Providence, especially Vice President David Brownell, and Firefighters Local 799, under the leadership of Leo Miller and Richard Kless, rescued our manuscript from the flames by providing timely grants to finance its publication.

In preparing the 2001 revision, we are indebted to Chief James Rattigan, Assistant Chief William Giannini, George Farrell, president of Firefighters Local 799, and Stephen Day, past president of that union for their insights and counsel. In addition, Michael Delaney, managing editor of photography at the *Providence Journal*, furnished valuable photographs, Linda M. Gallen prepared the revisions and final chapter for printing, and Anna Loiselle directed the marketing of the volume.

The authors hope this volume is worthy of all their efforts.

Patrick T. Conley
Paul R. Campbell

The first chiefs to figure prominently in this story were not fire chiefs but Indians. King Philip (shown here) was chief of the Wampanoag tribe. Greedy Plymouth colonists had forced him to make a number of humiliating concessions diminishing his power and his tribal lands. After a series of incidents in early 1675, war erupted with Philip's raid on Swansea (June 20, 1675). Eventually all of southeastern New England became involved in the bloody fray as Nipmucks, Narragansetts, and other tribes took the warpath.

On June 28, 1675, and again on March 29, 1676, Providence was attacked. The first raid, in which eighteen outlying houses were burned, was ordered by Philip (or Metacomet, his Indian name). The second foray was led by Narragansett chief Canonchet, who thereby obtained a measure of revenge for the December 1675 sneak attack by Plymouth soldiers upon his tribe at its village in the Great Swamp. More than two dozen houses near the heart of the small Providence settlement were torched in the second raid.

The war subsided in the late summer of 1676 after both Canonchet and Philip were killed. This conflict, which destroyed most of Providence's original homes, broke the power of the Indians of southern New England and took the lives of three thousand natives and six hundred English colonists. Wood engraving from History of King Philip's War, by Thomas Church, courtesy of Rhode Island Historical Society, Rhi x3 771.

chapter ONE

The Volunteer Era, 1754-1854

Roger Williams, leading a small band of religious dissenters, established the town of Providence sometime in the spring of 1636. Williams and his associates regarded their new settlement as "a lively experiment" in religious liberty and separation of church and state.

Although an embryonic police force, consisting of five "disposers" or arbitrators, was established in 1640, Providence had no fire officials of comparable authority until 1759, nearly a century and a quarter after the town's founding.

During that long period, however, a number of statutes were passed directed against the menace of fire. In 1647, for example, Rhode Island's first arson law was enacted as part of a comprehensive criminal code. It provided that "the penalty for burning dwelling houses or barns having corn in them, willfully and maliciously, is determined to be a felony of death without remedy." First-degree arson remained a capital crime until 1852, when the death penalty itself was abolished.

An October 1704 statute was Rhode Island's first fire prevention law. It banned the setting of fire "in the woods in any part of this colony on any time of the year, save between the tenth of March and the tenth of May annually nor on the first or seventh day of any week," under penalty of a thirty shilling fine. At this time Providence had the largest wooded area of any Rhode Island town, because its territory included all of present-day Providence County west of the Blackstone and Seekonk rivers (about 380 square miles).

Another fire prevention measure of 1731 banned unauthorized bonfires. Such blazes, in the years before the Fourth of July became their occasion, were usually set on Guy Fawkes Day, November 5, in celebration of the discovery and suppression of the Gunpowder Plot, a conspiracy in 1605 by a band of English Roman Catholics to blow up the House of Lords.

The statutes mentioned thus far were general laws applying to all Rhode Island towns. The first special act of importance pertaining to fire safety in a specific municipality was a 1698 law to prevent fires in Newport, the largest and most densely populated town in colonial Rhode Island. This measure required each of the householders in the heavily settled area near the harbor to "maintain one ladder," to clear his chimney regularly, and to refrain from littering the streets with straw, hay, or other "combustible matter." A second, more detailed law of May 1750 provided for the appointment of town officials called "firewards" to direct firefighting operations in Newport.

These fire laws, limited in scope and widely scattered in time, were the major measures enacted prior to the 1750s, when the townspeople of Providence took their first halting steps to establish a volunteer fire department. In October 1754, at the request of the Providence Town Meeting, the General Assembly appointed James Angell and merchant Obadiah Brown, uncle and mentor of the famous Brown brothers, to evaluate and to assess a tax on the housing and other property "in the compact part of the town of Providence which are liable to be destroyed by fire" in order to raise money to pay for a large water engine. The "compact part of the town," as it was called, extended along present North and South Main streets from Branch Avenue to India Point and also included Weybosset Neck (the present Downtown from Cathedral Square to the Providence River).

This first engine, manufactured by the highly regarded company of Newsham and Ragg of London, probably arrived in Providence sometime during 1755 and was housed on the gangway near the present junction of South and North Main streets.

The same act of the legislature that authorized "the speedy purchasing of said engine" also directed every Providence homeowner in the designed area to acquire, within three months' time, "two good leather buckets" to carry water to the new firefighting apparatus. In effect, every Providence housekeeper was recruited into the bucket brigade.

Often decisive actions, such as those fire safety measures implemented in 1754, come in the aftermath of tragedy, but other than the deliberate burning of numerous houses in 1675 and 1676 during King Philip's War, no major fires were recorded in Providence until just after the engine was placed in service. In 1758 the twenty-seven-year-old County House on the north side of Meeting Street burned to the ground. This building, where both the colony's legislature and the Town Meeting convened, was replaced in 1762. The destruction of the County House prompted the citizens of Providence to take additional steps to cope with the menace of fire.

In 1759 the Town Meeting purchased a second and larger Newsham engine and petitioned the General Assembly to pass "an act providing in case of fire breaking out in the town of Providence, for the more speedy extinguishing the same, and preserving goods endangered thereby." This law, based on the Newport statute of 1750, created two classes of fire officials appointed by the

Whereas the Inhabitants of the compact Part of the Town of *Providence*, represented unto this Assembly, That they conceive there is a great Necessity to have a Water-Engine of a large Size purchased to extinguish Fires that may casually break out in said Town ; and that the best Way to obtain one, will be by laying a Tax on the Houses, Goods, and other Things, liable to be destroyed by Fire.

On Consideration whereof,

BE it Enacted by this Assembly, and by the Authority of the same, It is Enacted, That *Obadiah Brown* and *James Angel*, Esqrs. be, and they are hereby impowered to rate the Housing, and all other Things within the compact Part of the said Town of *Providence*, which are liable to be destroyed by Fire, a Sum of Money sufficient to purchase an Engine as is above described ; and that the said Rate be levied so soon as may be, and the Money thereby raised, immediately put into the Hands of the said *Obadiah Brown* and *James Angel*, for the speedy purchasing said Engine.

AND be it farther Enacted by the Authority aforesaid, That every House-keeper, within the Space of three Months, be provided with two good Leathern Buckets, under the Penalty of forfeiting the Sum of Twenty Pounds ; one Half to the Informer, and the other Half to and for the Use of the Poor of the said Town of *Providence*.

AND be it farther Enacted by the Authority aforesaid, That the Money so raised, be by said *Brown* and *Angel*, subject to, and put under the Direction of the major Part of the Voters, among the Inhabitants of the compact Part of the said Town of *Providence*, and appropriated to purchase said Engine, in and after such Manner as they shall think fit.

This Assembly do Vote and Resolve, and it is Voted and Resolved, That *James Sheffield*, Esq; and Mr. *William Read*, be, and they are hereby constituted a Committee, to repair Fort *George*, and purchase Materials therefor.

This October 1754 statute, calling for the levying of a tax to purchase a fire engine for Providence, set in motion a chain of events that led to the creation of the Providence Fire Department. Photo of the original Schedules [Laws] of the General Assembly, October session, 1754, courtesy of Rhode Island Historical Society, Rhi x3 5002.

Town Meeting. One category, eventually called "presidents of the firewards," consisted of three (or more) "proper persons on whose fidelity, judgement and impartiality" the townspeople could rely. The presidents' functions were mainly concerned with the formulation of policy, but when fire did erupt, they were charged with giving "directions for the pulling down or blowing up of any such house or houses as shall be by them judged meet and necessary to be pulled down or blown up for preventing the further spreading of the fire."

The second category of officers, called "fire constables" and later "firewards," actually directed the firefighting efforts. They were empowered "to require and command allegiance for suppressing and extinguishing the fire, for removing household stuff and furniture... and for pulling down or blowing up" those houses ordered demolished by the presidents. The fireward's badge of authority and leadership, said the statute, was "a speaking trumpet, painted red." Disobedience to a fireward's commands brought heavy fines.

When the second Newsham engine arrived in 1760 and a third in 1763, the Providence Town Meeting decided to appoint volunteer "enginemen" to operate these machines. Since the original piece of apparatus was discarded as unserviceable, only two engine companies were formed. Company No. 1, consisting of eighteen men, was stationed near Market Square and operated the 1760 "downtown engine," while Company No. 2 (later called Niagara) was stationed near the foot of Smith Street and ran the 1763 "uptown engine." These appointments marked the culmination of a decade-long process by which the Providence Fire Department was formed.

The presidents and the firewards chosen by the Town Meeting were among the most prominent members of the community. The first group of three presidents appointed in 1759 included Esek Hopkins, who later became the first commander in chief of the Continental Navy. His brother Stephen—Rhode Island governor, chief justice of the state Supreme Court, and signer of the Declaration of Independence—was a president from 1768 to 1774, when his duties in the Continental Congress required his presence in Philadelphia. Nicholas Cooke, Rhode Island's Revolutionary War governor, served as president in the Providence fire service from 1762 through 1767. The most prominent of the firewards was prosperous merchant Nicholas Brown, who held that demanding post from its inception in 1759 until his death in 1791.

By 1771, as a result of numerous subdivisions of its territory, Providence had shrunk from about 380 square miles in land area to less than 6. Within its relatively narrow confines were approximately four thousand residents. At this juncture the Town Meeting, upon the recommendation of the presidents and firewards, promulgated "ten rules and regulations to be observed by the inhabitants in cases of fire." Some of these commands today seem curious and quaint. "If fire be cried at night," said one, "let every family immediately put candles in their windows next to the street" to light the way. Also, "Let every person, before he runs to a fire, take care to put on his clothes and take his buckets in his hand." In more substantive areas, these rules called for the creation of both a building committee, consisting of ten house carpenters, to pull down any buildings designated for demolition by the presidents to slow a fire's course, and a goods committee made up of "six or more elderly men, past hard labor," to pack up and remove goods from buildings threatened by fire.

In the following year (1772) a third engine company was formed on the west side of the river in a spot on Weybosset Neck called Muddy Dock, near the southerly end of Dorrance Street. The apparatus for this new unit (later called the Union Company) was manufactured in Providence by a craftsman named Daniel Jackson.

As the political crisis with England deepened, Providence citizens played a leading role in resisting the measures adopted by the mother country to compel the obedience of her American colonists. Two defiant acts of protest—both involving fire—were staged by the townspeople of Providence in the years prior to the outbreak of the Revolution.

In June 1772 a band of Providence merchants and mariners led by Captain Abraham Whipple and John Brown (a president of the firewards from 1785 to 1803) rowed down the bay to Namquid Point in Warwick and there boarded and burned the stranded British revenue sloop *Gaspee*, setting a fire that sparked the outburst of the

Providence's first three engines (one of which is shown here) were manufactured by the London firm of Newsham and Ragg, whose machines came in six sizes. Richard Newsham's primitive-looking pump was superior to any other eighteenth-century fire engine, but according to fire historian Paul C. Ditzel, credit for its development must be given also "to a mechanical genius of ancient Alexandria, named Ctesibius." Newsham's instrument was an improvement on that of Ctesibius, which was developed about 200 B.C. The Greek reputedly discovered the application of air as a motivating force in propelling water.

Ctesibius's device was a two-cylinder single-action pump shaped like an upside-down U. Forming the top of the U was a pivoting crossbeam, or rocker, at the ends of which were attached the pump handles. The rocker was connected to the pistons, or legs, of the U. Both legs, each snugly fitted into a brass cylinder casing, stood in a tub of water. On the upstroke of the pump handle one of the legs rose from the cylinder, and atmospheric pressure caused the cylinder to fill with water. On the downstroke the water was driven from the cylinder and into an air chamber, or dome. The purpose of the dome was to smooth out the pulsations to create a steadier water flow. The churning of the pistons up and down—alternately filling and emptying the cylinder casings—sent a continuous stream of water surging through the device, into the

dome, and out the discharge opening.

Newsham's device differed from Ctesibius's chiefly in its wheeled mobility. Otherwise, refinements were mainly cosmetic. The air chamber was concealed by a wooden housing at one end of the rig. Mounted atop this housing was a long, swiveling gooseneck nozzle that connected to the discharge opening. The Ctesibius pump

was the basic principle behind every fire engine for more than two thousand years, until around 1920, when the American fire apparatus industry ceased the manufacture of hand pumps. Photograph, courtesy of Rhode Island Historical Society, Rhi x3 1432.

From 1754 until the universal adoption of suction engines a century later, Providence households were required to maintain in readiness two leather buckets such as those shown here. These buckets, fashioned by shoemakers, were of two-gallon capacity. Townspeople formed bucket brigades to carry water from the nearest source to the engine box to be pumped onto the flames. Fortunately for Providence, many of its early structures were built on the cove or near the banks of the Moshassuck, Woonasquatucket, or Great Salt (Providence) River. Photograph by Denise Bastien.

final and decisive phase of opposition to England that preceded the armed rebellion. Then, on March 2, 1775, within the shadow of Engine 1, Providence rebels protesting the Tea Act of 1773 put the torch to three hundred pounds of the controversial leaf in Market Square. Boston had remonstrated against the measure with water; Providence chose fire to express its disapproval of the tax on what the town crier called "a needless herb, which for a long time hath been highly detrimental to our liberty, interest, and health."

During the American Revolution, Providence, sheltered at the head of Narragansett Bay, was spared occupation and the ravages of battle. Exposed Newport endured both, and so at war's end Providence emerged as the new state's major municipality.

During the decade of the 1780s, at least two private societies were formed in Providence for the mutual assistance of their members in time of fire. These auxiliary groups, the Amicable Fire Society (1785) and the United Fire Society (1786), were composed of local business and civic leaders. One of their major concerns was the protection of personal property within those buildings threatened by fire. In this mission they supplemented the efforts of the "goods committee" that had been established by the fire regulations of 1771 and became the forerunners of such official (i.e., public) agencies as the Furniture and Goods Protecting Company (1826) and the very useful and highly visible Providence Protective Company, created in 1875 by local insurance firms as an integral part of the fire department.

The bylaws of Amicable and other fire societies indicated that they were formed for social as well as utilitarian purposes. Three of the Amicable's charter members were governors of the state, and membership was limited to twenty-five persons. Article IX of its bylaws called for "a moderate supper provided quarterly" for those who belonged to this exclusive group.

As Providence continued to grow (population, 6,380 in 1790), more engines and enginemen were added to its fire service. A fourth machine, also built by Daniel Jackson, was purchased in 1791 and assigned to Benefit Street near Wickenden. The company appointed by the Town Meeting to man it eventually took the name Gazelle. A description

in the town records of the shed built to garage the engine indicates the modest construction of the early fire stations—thirty-four feet long, nine feet wide, thirteen feet high, and made entirely of wood. In 1798 the town got its fifth engine and the West Side its second when the Phenix Engine Company was formed. This twelve-man unit, using a machine built by Samuel Hamlin of Providence, occupied an enginehouse near Hoyle Square. Later it moved to Summer Street and adopted the new name Fire King.

At the dawn of the nineteenth century, Providence (population, 7,614) had a fire service administered by three presidents, seventeen firewards, about sixty enginemen, a five-member goods committee, and a nine-member building-removal committee. Despite these preparations, however, disaster struck at midmorning on January 21, 1801. Both in extent and in proportionate damage, the Great Blaze of 1801 was the most devastating in Providence history. The fire broke out in the loft of a large brick waterfront warehouse owned by John Corliss, a member of the Amicable Fire Society. Whipped by a cold, strong wind blowing out of the northwest, the flames raged furiously until midafternoon, when the destruction of several buildings, under order from the presidents, checked their spread. By the time the fire died out on the following day, it had cut a charred swath along South Main Street from Planet Street on the north to Williams on the south. No lives were lost, but thirty-seven buildings were destroyed and approximately $300,000 in property damage was inflicted, a sum equal to nearly 10 percent of the town's taxable property.

In the aftermath and as a consequence of the Great Blaze of 1801, the fire department underwent reorganization, and the Town Meeting promulgated rules and regulations to provide greater order and system to the task of firefighting. The presidents were clearly given the "supreme executive over all the fire officers" and were to carry a trumpet painted white as a symbol of their authority. Firewards were to direct on-the-scene operations at fires, issuing orders through trumpets of red. Two firewards were to supervise each of the engine companies, which were to be led by captains elected by the enginemen. The membership of the building committee

From the beginning, the Providence Fire Department was administered by community leaders of great prominence. Two such men were the Hopkins brothers, Esek (left) and Stephen. The former was one of the first three persons to be designated as "president of the firewards." Esek served in this capacity from 1759 through 1761. During the American Revolution this pioneer Providence fire official became the first commander in chief of the United States Navy.

Brother Stephen was a president of the fledgling department from 1768 to 1775, when his duties as Rhode Island delegate to the Continental Congress absorbed his full attention. In addition to his service as ten-time governor of Rhode Island and chief justice of the colony's Supreme Court, Stephen was a signer of the Declaration of Independence. Engraving of Esek by Dupin, from the author's collection. Portrait of Stephen by John Trumbull, from the author's collection.

The first comprehensive set of fire regulations for Providence was approved at an April 1771 town meeting and circulated to every householder by means of this handbill. The printer, John Carter, was editor of the town's first newspaper, the Providence Gazette and Country Journal. Broadside, 1771, courtesy of the Providence Public Library.

At a Town-Meeting held at *Providence*, the Seventeenth Day of *April*, 1771, the Committee appointed to draw up such RULES and REGULATIONS as are necessary to be observed by the Inhabitants in Cases of FIRE, reported the following, which were unanimously voted and agreed to by the Town, viz.

I. THAT upon the Cry of Fire, every Person take Care, at the same Time, to inform where the Fire is.

II. If Fire breaks in the Night, let every Family immediately put Candles in their Windows next the Street.

III. When Fire is cried, let the Engine-Men immediately repair to the respective Engines to which they belong, and let two of them, at least, tarry at the Engine-House till the Engine be gone; and then take Care that the Pipe, Hose, Buckets, and every Part of the Apparatus, be carried along.

IV. Let six proper Persons be appointed, whose Duty it shall be, upon the Cry of Fire, to repair to the Place where Fire-Hooks, Ladders, Ropes, &c. are kept, and to take Care that every Part of the Apparatus of that Kind be carried to the Fire.

V. Let every Person, before he runs to a Fire, take Care to put on his Cloaths, and take his Buckets in his Hand.

VI. When the People are assembled at a Fire, let them be as silent as possible, that they may hear the Directions of those whose Right it is to give Orders, and let them be executed with the utmost Alacrity, without Noise or Contradiction.

VII. Let the Presidents, and others who have Right to command at Fires, take great Care to appear calm and firm on those Occasions, and give their Orders and Directions with distinct Clearness, and great Authority; and be very careful not to contradict one another.

VIII. And let none vainly imagine such great Authority is given to the Presidents, Fire-Wards, and others, in the Case of Fire, meerly that they may command and domineer over their Neighbours; this is not the Reason of it, but the absolute Necessity of the Case requires it, and the Safety of the whole depend upon it; and therefore it ought to be chearfully submitted to, and willingly obeyed on these extraordinary Occasions.

IX. That the Town shall appoint ten House-Carpenters, whose Business it shall be, to remove or pull down any Houses or Buildings ordered by the Presidents; which Carpenters shall make necessary Rules among themselves, and appoint one to be their Chief; all which shall be observed and obeyed in Time of Fire.

X. That the Town appoint six or more elderly Men, past hard Labour, at a Fire, whose Business shall be, to give Orders for removing of Goods in Time of Fire, and whither they shall be carried; and every Person may pack up their Goods in order to be removed, but none of them shall be carried out (except of Houses actually on fire) but by Order of the said Persons, or some one of them, who shall be careful to give their Orders seasonably, so that no Goods be lost that can be removed.

XI. That a sufficient Number of these Rules and Regulations be printed, and every House-keeper furnished with one of them.

STEPHEN HOPKINS,
JOSEPH BROWN,
WILLIAM SMITH, } COMMITTEE
BENJAMIN MAN,

April 17, 1771.

In June 1772 Providence citizens resorted to fire in expressing their opposition to British economic regulations. On June 9 the packet Hannah, *plying between New York and Providence under command of Captain Benjamin Lindsay, was chased up Narragansett Bay by the dreaded British customs sloop* Gaspee, *Lieutenant William Dudingston commanding. Attempting to evade capture, Lindsay sailed into shallow water off Warwick, and the Englishman, pursuing recklessly, ran his larger vessel aground on Namquid (now Gaspee) Point.*

Lindsay docked in Providence, where that evening the townsmen, led by merchant John Brown, assembled in Sabin's Tavern (on the corner of Planet and South Main streets) to plot the destruction of the Gaspee. *Discussions continued until 10:00 P.M. Then the rebels*

embarked from Fenner's wharf in eight 5-oared longboats under the command of Abraham Whipple, a highly successful privateer captain in the French and Indian War. After midnight the attack party reached the stranded ship, and following an exchange of shouts James Bucklin shot and wounded Lieutenant Dudingston in the groin. The Providence men thereupon boarded the Gaspee, *overpowered the crew, and burned the sloop and its contents.*

A royal investigation of the incident yielded insufficient evidence to indict the perpetrators, who were shielded by their fellow townsmen. But the Gaspee *inquiry led to the establishment of legislative committees of correspondence throughout the colonies, a major step on the road to the Revolution. Burning of the* Gaspee, *steel engraving by J. Rogers, courtesy of Rhode Island Historical Society, Rhi x3 119.*

In the closing years of the eighteenth century, merchant John Brown (whose mansion still stands at 52 Power Street) was Providence's most prominent citizen. His many civic functions included service as a president of the firewards from 1785 until his death in 1803. In a letter to his son written in March 1783, Brown gave a firsthand account (his words and his spelling) of a typical firefighting effort:

Yesterday morning Fire was Cried Uptown and the Church Bell Soon Began to Ring, the Whole Town as it Ware was Instantaniously In Motion I had not got far before I heard it said the Fire was Job Smiths. . . . I concluded to myself the House was gone, but soone Discovering their was no Fire in the middle of the House & that it was only in the Roofe from the Chimney to the South End, the water being handy with Three Injoines and a multitude of people we soone mastered the fire and saved the House. . . Water will dammage the paper and furniture some, their was but little of the household stuff removed which proved a happy surcumstance. Indeed, the lanes of men formed to supply the Ingoines with water surrounded the house so that it was difficult to get in or out.

Miniature by Edward G. Malbone, courtesy of New York Historical Society.

of house and ship carpenters was increased to thirty. To their duty of tearing down endangered buildings to check a fire's spread was added the duty of caring for "fire hooks, ropes, ladders, axes, crow-bars, and shovels" and conveying them to the fire. Their badge of office was designated "the leather cases of the axes painted white." The goods committee was increased to twelve, and each member was given a "white wand or staff of six feet" as an identifying symbol of authority. The rules also detailed the duties of the general public and required an annual firefighting drill or maneuver for all fire officials, with the townspeople welcome to attend "with their buckets." Finally, a system of alarm was established "whereby the several sextons repair immediately to their respective meetinghouses, and ring the bells until the fire be extinguished."

During the period prior to the formation of the Board of Firewards in 1825, the presidents and firewards met irregularly (and infrequently, if their surviving records are complete) to formulate fire policy. In 1805 this group required ladders and axes to be standard equipment for fire engines. In 1814 three wardens were assigned to supervise each engine company. In that year Samuel Bridgham, who was destined to become the city's first mayor, was named a fireward for Engine 1. During these years fire officials continued to be drawn from the civic leadership of Providence. They included Edward Carrington and Nicholas Brown (the younger), the city's two leading merchants; Samuel G. Arnold, the state's premier historian; and prominent businessmen such as Ebenezer Knight Dexter, Cyrus Butler, Jabez Bowen, James Fenner, Richard Jackson, John Dorrance, Elisha Dyer, Sr., Crawford Allen, John Howland, Sullivan Dorr, and Thomas C. Hoppin. Firefighting was a gentleman's duty and avocation.

From 1801 to 1814 Providence suffered no spectacular conflagrations. Then, in June 1814, an arsonist torched the architecturally ornate meetinghouse of the First Congregational Society, which stood at the southeast corner of Benefit and Benevolent streets. Pastor Henry Edes set the loss at forty thousand dollars and posted a five hundred dollar reward for the firebug, but to no avail.

The year 1814 is also significant for the formation of

the town's sixth engine company in the village of Olney-ville, which then rested on the old boundary between Providence and the adjacent towns of North Providence and Johnston. This unit (eventually called Eagle Engine Company for Olneyville's Eagle Mill) carried the number 6 until 1832. Then it assumed the title Engine 1 when that designation became available through dissolution of the 1763 Market Square, or "downtown," company.

Also in 1814 the civic demands placed upon the "bucket brigade" were increased. The penalty on each householder for failure to "keep two good leather buckets, containing at least two gallons each, with the owner's name painted at large thereon" was increased to five dollars (about an average week's pay), and the town sergeant and the constables were directed to conduct annual inspections to see that this law was observed.

As Providence grew more densely populated owing to the onset of industrialization and the subdivision of its once sizable real estate parcels, wooden buildings posed a serious fire hazard. A state statute passed in October 1817 at the request of the townspeople barred the future construction in the compact area of Providence of wooden buildings that were more than thirteen feet high from the ground to the peak of the roof. The district encom-passed by the preventive statute comprised the present Downtown and the western slope of College Hill from Wickenden to Smith Street. Heavy taxes and other penal-ties were imposed upon violators, and provisions were made for the expansion of the zone. In 1834 two-story wooden dwelling houses were exempted from this ban, and later the building height was increased to eighteen feet. This important statute, though modified by sub-sequent enactments, remained Providence's basic build-ing code until 1878.

The 1820s witnessed a marked increase in the size and scope of the Providence fire service. As the decade dawned, the town's first hook-and-ladder company was formed—the hooks to pull down threatened wooden structures and the ladders, of course, to scale the taller buildings that were becoming more commonplace in the rapidly growing town. Seven years later a second such unit was added to the town's fire service.

In 1807 two Philadelphia firemen, James Sellers and

Abraham L. Pennock, had devised a method for improv-ing the leather hose then in use by adding metal rivets to bind the seams. Until their innovation hose seams were merely stitched, and thus they leaked badly and failed to withstand the higher water pressure that the newer pumps could deliver. Sellers and Pennock formed a company to produce their hose and then turned to the manufacture of hose carriages and engines as well. Ac-cording to one leading authority, "the significance of the Sellers and Pennock hose for American firefighting can-not be overemphasized." The sturdier hose spurred the development of suction techniques to draw large quan-tities of water quickly and directly into the engine. The suction method—the wave of the future—would displace the slow and tedious process of dumping buckets of water into the engine box. Although the suction principle was known as early as 1698, implementation of the idea was delayed because suction required a stronger hose.

Sellers and Pennock not only resolved the hose problem; in 1822 they manufactured the first fire appara-tus with suction hose and fittings. This pioneering rig—an end-stroke pumper with a single piston and a one thousand-foot reel of hose—was bought by the town of Providence and named Hydraulion No. 1. The credit for procuring this revolutionary piece of equipment goes primarily to businessman and inventor Zachariah Allen and to Elisha Dyer, Sr. In January 1823 the new device was entrusted to the supervision of three firewardens—Ed-ward Carrington, the town's leading merchant; John H. Greene; and William Blodget. It was manned by a newly created volunteer company and housed in a wood build-ing erected on the shore of the old cove at the present northwest corner of Kennedy Plaza and Exchange Street (then named Hydraulion Street). This engine, which cost $725, was manned by thirty-six firefighters who worked its "brakes" (as the two horizontal pumping handles on either side were called). It could lift water from the cove and discharge a powerful stream more than one thou-sand feet from its source. It was estimated that these thirty-six men replaced about six hundred men, women, and children of the old-style bucket brigade. In 1831 the firewardens bought a second unit for the newly estab-lished Hydraulion Company No. 2 on Canal Street.

ARTICLES

OF THE

AMICABLE FIRE SOCIETY.

THE Members of the AMICABLE FIRE SOCIETY, repofing fpecial Truft and Confidence in each other, mutually agree to the following ARTICLES:——

I.

THAT whenever it fhall pleafe GOD to permit Fire to break out in the Town of *Providence*, we will then faithfully aid and affift each other, him firft who is in the moft apparent Danger.

II.

THAT each Member fhall furnifh himfelf with Two good Leather Buckets, Thirteen Inches deep, Ten Inches broad at the Top, and Seven and an Half Inches at the Bottom; and Two good Ravens-Duck Bags, One Yard and an Half long, and Three Quarters of a Yard wide, with drawing Strings: That the Buckets and Bags fhall have the Initials of the Chriftian, and the Whole of the Sirname of the Owner, with the Device of the Society (which is Two Rivers, blending in a common Stream) painted thereon: That each Member fhall keep his Bags in his Buckets, in the moft confpicuous and eafy Place to come at, in his Houfe: That they fhall not be ufed, but in cafe of Fire; and that all Buckets and Bags, damaged or loft at a Fire, where the Owner or any of his Family were prefent, fhall be made good by the Society.

III.

THAT if it fhould at any Time be deemed neceffary to remove the Goods or Effects of any Member out of his Houfe, Shop or Store, they fhall be removed by the Members of this Society, to as few Places as may be; and, when depofited, One or more Members fhall ftay by, and fee them fafely fecured.

IV.

THAT if the Houfe, Shop, Store, Goods or Effects, of any Member, fhall be in Danger, in his Abfence, the Members prefent fhall ufe their utmoft Exertions for the Prefervation and Security thereof; and that if the Houfes, Shops, Stores, Goods or Effects, of Two or more Members fhall be in Danger at the fame Time, the other Members fhall divide for their Affiftance.

V.

THAT no Member may be impofed on by evil-minded Perfons, there fhall be a WATCH-WORD agreed on at every Quarterly-Meeting, to be ufed when neceffary; and that it may not be forgot, every Member fhall repeat it to the Clerk, at the next Meeting.

VI.

THAT every Member fhall, if Health permit, attend all Fires that may happen in Town, and exert himfelf to extinguifh the fame, and affift the Sufferers, whether connected with this Society or not.

VII.

THAT this Society fhall at no Time confift of more than Twenty-five Members, Thirteen of whom are to compofe a Meeting to tranfact Bufinefs, and a Majority of their Votes to be binding on the Society: That the Society meet Four Times in a Year, at the Hour of Six o'Clock, P. M. in *January* and *April*, and at Seven o'Clock in *July* and *October*, on the Firft *Thurfday* in each of faid Months, at fuch Places as the Majority fhall determine; and that no Vote be paffed after Nine o'Clock in *January* and *April*, nor after Ten o'Clock in *July* and *October*. At each Meeting a Moderator fhall be chofen, that the Bufinefs may be conducted with Order and Decorum; and that at the *January* Meeting annually a Clerk fhall be chofen for the Year, who fhall keep a fair Record of the Proceedings of the Society, receive and be refponfible for all Fines, and at the End of the Year render an Account of the Difpofal thereof; and he fhall alfo every Quarter, at leaft Two Days before the Meeting, take with him Two Members by Rotation, and vifit the Habitation of each Member of this Society, in order to inform themfelves fully of their Situation,

and the moft eafy Way of coming at their Dwellings in cafe of Fire, and infpect their Buckets and Bags, and report their State to the Meeting; for which Services he fhall be exempted from his Proportion of the Expence of the Four Quarterly-Meetings.

VIII.

THAT whenever a Vacancy fhall happen in the Society, by Death or otherwife, the Perfon who is a Candidate fhall be propounded Three Months before the Ballots are taken for his Admiffion, which fhall be unanimous.

IX.

THAT there fhall be a *moderate* Supper provided quarterly, at the Expence of the Society, under the Direction of the Clerk, who fhall collect and pay the Reckoning.

X.

THAT if any Member fhall neglect or refufe to obferve any of the preceding Articles, he fhall pay the following Fines, affixed to the Offences, unlefs he render a fatisfactory Reafon therefor.

	Lawful Money.
For Non-Attendance on any Fire in Town,	£ 0 18 0
For not keeping Buckets and Bags in Repair, each	0 6 0
For ufing Buckets or Bags, except at a Fire, each	0 6 0
For not carrying Buckets and Bags to a Fire, each	0 6 0
For Non-Attendance on each Meeting,	0 3 0
For Non-Attendance, in One Hour after the Time affixed,	0 1 6
For refufing to ferve as Clerk, when chofen in Turn,	0 12 0
For omitting to bring thefe Articles to each Meeting,	0 1 6
For refufing to attend the Clerk, to vifit the Members, when called on by Rotation,	0 6 0
For forgetting the Watch-Word,	0 1 6

AND that any Member neglecting to attend Four Quarterly-Meetings, fucceffively, fhall be excluded the Society, and his Name erafed.

In Witnefs whereof, we have hereunto fet our Hands, in *Providence*, this Twenty-fecond Day of *December*, A. D. 1785, and of AMERICAN INDEPENDENCE the Tenth.

JEREMIAH OLNEY,
WILLIAM HOLROYD,
GEORGE OLNEY,
SAMUEL YOUNG,
GEORGE BENSON,
ANDREW DEXTER,
WILLIAM LARNED,
WILLIAM PECK,
CYPRIAN STERRY,
WILLIAM CORLIS,
EPHRAIM BOWEN, jun.
WILLIAM JONES,
JOHN DORRANCE,
THOMAS L. HALSEY,
OLNEY WINSOR,
NICHOLAS COOKE,
JOHN ROGERS,
LEWIS DEBLOIS, jun.
JEREMIAH F. JENKINS,
PARDON BOWEN,
WILLIAM A. SESSIONS,
SAMUEL SNOW,
STEPHEN HOPKINS,
JOHN YOUNG,
JOSEPH JENCKES.

PROVIDENCE: Printed by JOHN CARTER.

During the 1780s several private societies were formed in Providence and Newport for the mutual assistance of their members in time of fire. Such auxiliary fire organizations as the Amicable Fire Society had a limited membership consisting mainly of civic leaders and businessmen, and these groups served social as well as practical purposes. Among the Amicable's twenty-five members were three Rhode Island governors—Stephen Hopkins, William Jones, and Nicholas Cooke. Broadside, 1785, courtesy of Rhode Island Historical Society, Rhi x3 4962.

The early fire alarms were provided mainly by church bells. During the late eighteenth century the major "town bell" was located in the steeple of the First Baptist Meeting House, built in 1775. Other bells rang from the steeples of the original St. John's Church, on the North Main Street site of the present Cathedral of St. John the Divine, and the Congregational Meeting House, located at Rosemary Lane (College Street) and Back Street (Benefit). Steel engraving by J. W. Lincoln, G. G. Smith, and J. W. Watts, courtesy of Rhode Island Historical Society, Rhi x3 1196.

The second major codification of local fire regulations was issued on February 20, 1801, in the immediate aftermath of the disastrous blaze—Providence's worst—that destroyed thirty-seven buildings and nearly 10 percent of the town's taxable property. The thirteenth rule is an excellent description of the average citizen's role in combatting a major fire in early nineteenth-century Providence:

> When people begin to assemble at a fire, before the engines or any appointed authority arrive, they should not wait for orders, but immediately proceed to carry water from the nearest and most convenient place they know of to the fire, and as soon as more are assembled than can get convenient access to the fire, they should begin to form a lane from the fire towards the most convenient place for water, and from thence towards the fire. The youth who are not able to endure the fatigue of handling full buckets should all form on that side of the lane that brings their right hand towards the water and their left towards the fire, this being the side for returning the empty buckets, and where they may perform the service of men. When more water can be procured from the place where the first lane is formed than one row of buckets will convey, let a double lane be formed by adding a third row of men on the outside of the youths' row, or that which returns the empty buckets, and let every other person in the youths' row face about towards the new-formed row, that they may with more convenience pass the empty buckets to the water as fast as the two rows of full buckets require, until more people arrive to form another row. And as water is passed much easier, in buckets as well as hose, down-hill than up, care should be taken to bring it from higher ground when it can be got at nearly equal distances.

Broadside, 1801, courtesy of Rhode Island Historical Society, Rhi x3 4965.

The following are the RULES and REGULATIONS for the Government of the Inhabitants of the Town of PROVIDENCE, in Cases of FIRE, as reported by a Committee, and adopted by the Town, at their Meeting holden by Adjournment on the Fourteenth Day of February, A. D. 1801.

FIRST. THAT upon the cry of Fire, every person give information (if within his knowledge) *where the Fire is*; and that the several Sextons repair immediately to their respective meeting-houses, and ring the bells until the fire be extinguished.

SECOND. The Engine-Men, the two attending Fire-Wards, and the Watermen, should repair immediately to their respective engines, and conduct them with dispatch to the fire. The Wardens are to see that the pipe, suction, hose, buckets, copper pump, and all other apparatus thereto belonging, be forwarded with the engine.

THIRD. That all other able-bodied male inhabitants repair immediately, with the buckets belonging to their respective families, to the fire; taking care, if in the night, to put on their clothes before they go out; and every house should have lights put in the windows, and carefully attended until the fire is extinguished, and the people returning.

FOURTH. That the thirty house and ship-carpenters, annually appointed for taking down buildings, &c. have the care of the fire-hooks, ropes, ladders, axes, saws, crow-bars and shovels, and immediately convey them to the fire, and there exert themselves to extinguish the same, under the direction of the Presidents and Fire-Wards; and they are to meet the day after their appointment, and elect one to be their Chief, and such other officers as they may think proper; and to consult and agree on the best method to speedily convey the apparatus under their care to fires, and of using the same when there in the most effectual manner:—Their badge of office shall be the leather cases of the axes, painted white.

FIFTH. The Presidents should each have, and take with them, a trumpet painted white, and the Fire-Wards, each, one painted red, as their badge of office, and that they may be the better heard and understood. And let all who have a right to command at fires, take great care to appear calm and firm, and give their orders and directions with clearness and authority, and be careful not to contradict each other.

SIXTH. When the people are assembled at fires, let them be silent, that they may hear the directions of those whose right it is to give orders; and let them be executed with the utmost alacrity, without noise or contradiction.

SEVENTH. Let none vainly imagine that the great authority given to the Presidents, Fire-Wards, and others at fires, is, that they may domineer over their neighbours; this is not the case, but authority and order are absolutely necessary, and the safety of the whole thereon depends; and therefore ought to be cheerfully submitted to, and willingly obeyed, on these extraordinary occasions.

EIGHTH. The twelve suitable persons appointed for the removal of goods, shall each have authority to give orders, and command assistance for the removal, preservation and safe keeping of goods; and though the owners of goods may pack them up in order for removal, yet none of them should be ordered out, except of houses actually on fire, unless by the direction of the owner, or of some one of the said persons, who should be careful to give seasonable orders, that they may be carried a proper distance to *windward* of the fire, that no goods be lost which can be removed. The badge of this class of officers shall be a white wand or staff, of six feet, which they are to take with them.

NINTH. In point of authority and subordination, the Presidents of the Fire-Wards have the supreme executive over all the fire officers, and are to call upon whomsoever they judge proper, for aids, messengers or assistants, in time of fires. The Fire-Wards are to attend to the general subject of extinguishing and preventing the spreading of fire, direct the stands and operations of the engines, forming of lanes for conveying of water, further the exertions of every department, and encourage the citizens at large in active and persevering attention to the preservation of the lives and property of their fellow-citizens in immediate danger, and the general safety and interest of the town. They are to meet at the Council-Chamber the day after their appointment, annually, and afterwards as often as they may judge proper, to select two of their number to pay particular attention to each engine, and see that they are supplied with water, and the Engine-Men duly assisted with frequent, sufficient and fresh aid to work the engines, and to confer on the general subject of preserving the town from fires, and the best method of extinguishing them, in concert with the Presidents, who are requested to meet with them, and appoint such other times for general conference and consultation on this interesting subject as they may judge proper.

TENTH. The Engine-Men are to keep near the engines; to be always in readiness, under the direction of their respective captains, whom they shall appoint, to remove the engines from place to place; to work them with or without fuction, with or without hose; to convey water from the river, fountains or wells, to other engines; to give place to or exchange stands with other engines, as the Fire-Wards may direct; and at all times to keep the engines and apparatus in good order, fit for use, at the town's expence; and the captains are to name, at every fire, one or more of their respective companies to have particular care of the hose, fuction and apparatus immediately connected with the engines; to see that they are kept in safety from the fire, and ready for instant use.

ELEVENTH. That the Town-Sergeant, for the time being, cause all the buckets left, after every fire shall be extinguished, to be forthwith carried at the town's expence to the Market-House, before night, if there be time, if not, the next day.

TWELFTH. The Presidents of the Fire-Wards, as often as they judge proper, at least once a year, are to give public notice to all the fire officers under appointment by the town, and all other inhabitants that are free to attend with their buckets, to punctually collect the several engines, at the time and place they shall assign, in order to go through all the necessary manoeuvres usually required with the apparatus used in extinguishing fires, for their own improvement by experience, and for the instruction of the rising generation.

THIRTEENTH. When people begin to assemble at a fire, before the engines or any appointed authority arrive, they should not wait for orders, but immediately proceed to carry water from the nearest and most convenient place they know of, to the fire; and as soon as more are assembled than can get convenient access to the fire, they should begin to form a lane from the fire towards the most convenient place for water, and from thence towards the fire. The youth who are not able to endure the fatigue of handing full buckets, should all form on that side of the lane that brings their right hand towards the water, and their left towards the fire, this being the side for returning the empty buckets, and where they may perform the service of men. When more water can be procured from the place where the first lane is formed than one row of buckets will convey, let a double lane be formed, by adding a third row of men on the outside of the youth's row, or that which returns the empty buckets, and let every other person in the youth's row face about towards the new-formed row, that they may with more convenience pass the empty buckets to the water as fast as the two rows of full buckets require, until more people arrive to form another row. And as water is passed much easier, in buckets as well as hose, down hill than up, care should be taken to bring it from higher ground, when it can be got at nearly equal distances.

A true Copy: Witness,

NATHAN W. JACKSON, *Town-Clerk.*

Providence, February 20, 1801.

PRINTED BY JOHN CARTER.

The second meetinghouse of Providence's First
Congregational Society was built under the
direction of pastor Enos Hitchcock in 1795 at
the southeast corner of Benefit and Benevolent
streets. This ornate twin-spired church was
destroyed by an arsonist in June 1814, causing
a loss estimated at forty thousand dollars. The
present structure, still standing as a Unitarian
house of worship, replaced it in 1816. To the
right in this painting are two of Providence's
most impressive Georgian brick structures, the
Thomas Poynton Ives House (1806) and the
John Brown House (1786), extreme right, with
its four outbuildings. Anonymous painting,
circa 1810, courtesy of Rhode Island Historical
Society, Rhi x3 3036.

During this period yet another innovation helped effect the demise of the bucket system—the stationary forcing engines. These devices were designed and built by local machinists Albert H. Manchester and John T. Jackson and were operated by manual power applied to brakes. They were "stationed" at wells or cisterns strategically located throughout the town. The first of fifteen engines or pumps that were built between 1825 and 1834 was located near present-day Weybosset and Mathewson streets. Those units on the east side of the river were maintained by a newly formed company called Canopus No. 1, located on Benefit near College Street. Invincible No. 2, captained by Jacob Manchester, ran the west side pumps from a station located on Middle Street near Dorrance. Eventually reaching sixty-nine in number, these units became obsolete with the introduction of improved water service and were gradually phased out in the late 1880s.

In addition to the hydraulion, two stationary engine units, and two hook-and-ladder companies, one new engine company made its debut in the 1820s, Engine 7. This group, eventually called the Blue Pointers or the Ocean Company, was located first on Chestnut Street (1824), then on Field (1832), and finally, by 1839, on Richmond Street.

An 1822 town ordinance required each fireman to wear "a uniform badge bearing the number of the engine to which he is attached, that such may be distinguished from other citizens in time of fire." Henceforth every appointed member of the Providence fire service—from president to engineman—had his special insignia.

Other fire laws during this decade included a safety measure passed on the eve of America's fiftieth anniversary of independence in anticipation of what threatened to be a wild and riotous celebration. The legislature enacted a fireworks statute forbidding the sale or use of "crackers, squibs or other fireworks of a combustible nature" without a special license from the town council. Other notable measures included an ordinance (1823) prohibiting "the smoking or carrying of any lighted cigar or pipe" in any of the streets or on any wharf in the compact part of town; an act (1825) allowing the firewards (without Town Meeting approval) "to purchase such new

engines and necessary apparatus as they may think proper and expedient for the use of the town"; and a law (1827) forever exempting firemen from military duty, except in time of war, "provided they shall serve in said capacity ten years in succession, after they arrive at the age of twenty-one." This statute was designed as an inducement for young men to enter the fire service.

To insure greater discipline in the department, a state statute was passed in November 1826 giving fire companies power to enact bylaws and to inflict fines and penalties on their members. In furtherance of this statute, one volunteer company enacted a regulation stating that "no cause for non-attendance shall be received except a certificate from a physician or undertaker." The 1826 law was the forerunner of several special acts passed in the 1830s and 1840s conferring corporate charters (and thus greater autonomy) upon the fire companies of Providence.

In 1825 the presidents and the firewards adopted a more formal mode of organization. They organized themselves into a Board of Firewards, elected a chairman (the longest tenured of whom was Ephraim Talbot, 1829-1835), and began to hold frequent and regular meetings to formulate department policy. For more than two decades (until 1853) this board ran the department, buying engines, hose, and other apparatus, selling old engines, equipping the force with "India rubber" overcoats, building stations, preparing budgets, framing regulations, enforcing the building code of 1817, acquiring sites for wells, and directing firefighting operations. Its chairman was the de facto chief of the fire service.

The last important event of this active decade was the incorporation of the Providence Association of Firemen for Mutual Assistance. This organization was formed by Zachariah Allen, Amasa Manton, and others for the purpose of providing aid to firefighters injured in the performance of their duties and benefits to the families of those who lost their lives. The death of Joshua Weaver while combatting a South Main Street blaze on March 20, 1828, was the catalyst for the creation of this private relief agency, to which both the individual firemen and the town contributed annual stipends.

In 1832, following a bloody riot in September of the

The pride of the Providence fire service in the volunteer era was Hydraulions No. 1 (1822), above, and No. 2 (1831), manufactured by Sellers and Pennock of Philadelphia. The 1822 hydraulion (from the Greek word meaning "water") was the first piece of fire apparatus with suction hose and fittings to be put in use anywhere. The suction method of drawing large quantities of water directly into the engine rapidly displaced the traditional bucket brigade.

When Hydraulion No. 1 arrived, the leading citizens of the town displayed an eagerness to form a company to run it. No less a man than Edward Carrington, one of Providence's wealthiest merchants, was elected foreman of the forty-five member Hydraulion Company No. 1, which also included Zachariah Allen, Moses Brown Ives, Robert H. Ives, Amasa Manton, and Nehemiah Dodge, the noted silversmith.

Technological change was so rapid in the fire service that these revolutionary machines became obsolete by the 1840s. Both were unceremoniously sold in 1854 to unknown buyers, probably mills. Their ultimate fate has never been discovered. Engraving from The Providence Plantations for 250 Years, by Welcome A. Greene, courtesy of Rhode Island Historical Society, Rhi x3 4946. Painting by Charles A. Foster, 1850, courtesy of Rhode Island Historical Society, Rhi x3 3103.

Zachariah Allen exerted a great impact on fire-fighting efforts in nineteenth-century Providence. Allen (1795-1882) was an author, lawyer, medical doctor, town councilman, historian, and social reformer, as well as a first-rate scientist and inventor. In 1822 he introduced the town's original hydraulion (suction fire engine and hose equipment) to replace the hand-buckets previously used. Later he became a principal developer of the Providence waterworks. Among his many industrial patents were a central furnace system for heating houses by hot air, a method of transmitting power by leather belting in place of the gear or "cogwheel" connections previously employed, a cloth-napping machine, a dressing and finishing machine for cotton textiles, a machine for spooling wool, and the automatic steam-engine cutoff valve.

Allen also originated a system of mutual fire insurance for manufacturing property that required underwriters to study methods for fire prevention and to calculate premiums on the adequacy of the safety equipment installed. To implement this approach, he founded the Manufacturers Mutual Fire Insurance Company in 1835, the flagship of the Factory Mutual System and the forerunner of the prestigious Allendale Mutual Insurance Company. Allen was also a founder and trustee of the Providence Association of Firemen for Mutual Assistance (1829), which provided assistance to families of firemen injured or killed in the line of duty. Photograph, courtesy of Rhode Island Historical Society, Rhi x3 4275.

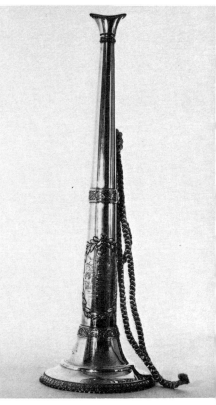

Typical of early nineteenth-century firefighting equipment were leather buckets, salvage bags (this one bearing the name of Zachariah Allen), and the indispensable fireman's trumpet. Photographs by Denise Bastien, courtesy of Museum of Rhode Island History, Rhode Island Historical Society.

previous year, Providence abandoned its town meeting form of government and became a city with a mayor-council system. At this juncture the annual report of the firewards indicated the status of the fire department: 5 presidents, 26 firewards, 500 firemen (nearly a fivefold increase since 1822), 10 station houses, 7 engine companies, 8 hand engines, 12 forcing stationary engines, 2 forcing stationary engine companies, 2 hook-and-ladder companies, 2 hydraulions and hydraulion companies, 2 hose wagons, and 4,955 feet of hose. The department was taking shape as one of the nation's most advanced.

Three years later, in 1835, the firewards proudly reported the firefighting prowess of the department was such that a relay of forcing pumps and hydraulion had conveyed water a distance of twenty-five hundred feet and supplied an engine in thirty minutes from the first alarm. At that time the capacity for throwing water on a fire in the compact portion of the city was eight hundred gallons per minute; in more remote sections, five hundred gallons per minute; and in the "suburbs," three hundred gallons.

In 1835, the year of the first annual firemen's parade, an event of great significance occurred in Providence when Zachariah Allen, upset by the cost of conventional insurance for his Allendale woolen mill, founded a mutual fire insurance company devoted exclusively to insuring factories against fire loss. This business, called the Manufacturers' Mutual Fire Insurance Company, was the pioneer in the now famous Factory Mutual System. It was Allen who first devised a procedure for inspecting and rating factory buildings and setting their premiums based upon their condition, their method of construction, and the quality of the fire prevention apparatus they installed. His formula drastically lowered fire insurance rates for those industries that practiced fire safety. As one historian of the insurance industry has obvserved, "Allen took the first step toward adapting the mutual idea to meet the needs of the business community and began the integration of mutual insurance into the commercial life of the nation."

During this era fire insurance also became a mainstay of the city's oldest and largest insurance company—Providence Washington. Early in the presidency of Sulli-

van Dorr (1838-1858), this firm, in whose conference rooms the Board of Firewards often met, temporarily phased out the marine insurance on which it was founded in 1799 and turned its full energies to the field of fire coverage. A third major Providence company to underwrite this type of insurance was Firemen's Mutual, established in 1854. This firm and Manufacturers' Mutual are direct lineal ancestors of today's local giant, the Allendale Mutual Insurance Company of Johnston.

In the twenty-two years between the incorporation of the city and the establishment of a paid department, five additional engine companies were formed and several existing companies were chartered and enlarged by special act of the General Assembly.

The dissolution of Engine No. 1 in 1831 after sixty-eight years of service caused considerable confusion in numbering the city's several volunteer units. For example, Engine 6 of Olneyville (established 1814) earned the designation of Engine No. 1 in the following year because of its proficiency, leaving the Water Witch Company, founded in 1834, to assume the title Engine No. 6. Water Witch, stationed on the town lot at the corner of Benefit and College streets, was equipped with a highly regarded gooseneck pumper built by James Smith of New York. Water Witch was the first company to secure a corporate charter from the General Assembly (1834) and the first to adopt a uniform (1843). Among its members were Captain Joseph W. Taylor, who later became the paid department's first chief, and Edward Lewis Peckham, Providence's most talented landscape painter.

In 1836 Pioneer Engine Company No. 8 was chartered. It took up its initial quarters at the corner of Benefit and Transit streets in a substandard structure that burned down in December 1837, to the embarrassment (or, perhaps, satisfaction) of the contentious Pioneers. Shortly after this suspicious fire, the company secured both a new Smith engine and a new station house on the same site. The free-spirited Pioneers operated with efficiency and zeal until 1846, when they got into a row with the firewards over the failure of the latter to provide the company with improved "New York" hose of greater capacity. In October 1846, though the Pioneers (as usual) had raced first to the scene of a fire, they were "washed out" by another

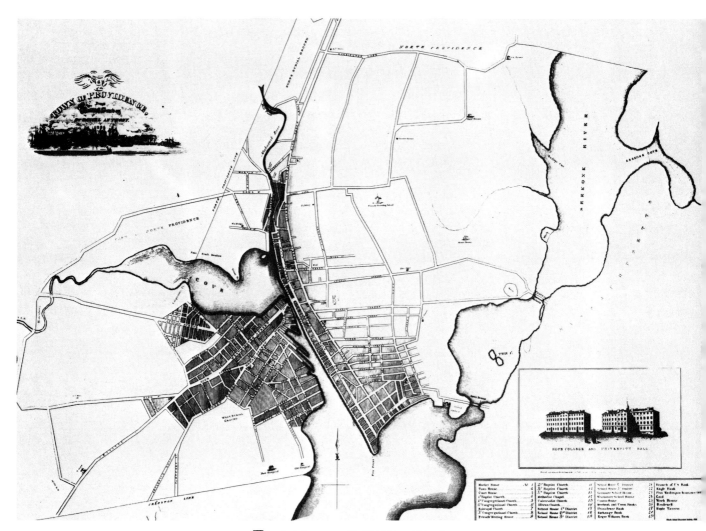

The shaded area on this 1823 Daniel Anthony
map corresponds to "the compact part of town"
to which the early fire laws applied. Note the
existence of a large cove on the site of the present
Capital Center Project. The North Providence
town line ran near the present junction of Orms
Street and Douglas Avenue and thence westerly
along the Woonasquatucket to Olneyville, while
the Cranston town line was located just south of
the hospital near present-day Dudley Street.
Early nineteenth-century Providence was less
than six square miles in area—about 30 percent of its present size. Map by Daniel Anthony,
1823, courtesy of Rhode Island Historical
Society.

$300 Reward!

Satisfactory information having been given to this board that an attempt was made last evening to burn the Meeting-House heretofore belonging to the Pacific Congregational Society, standing at the southerly corner of Richmond and Pine streets, by kindling a fire in the cellar thereof, which had it not been timely discovered, would have enveloped said building in flames,

It is therefore, Resolved, That a reward of three hundred dollars be paid for the discovery and apprehension of the person or persons who shall be convicted of said offence.

City Clerk's Office,
Providence, Feb. 24, 1834.
I certify that the above is a true copy of a resolution this day passed by the Board of Aldermen of the City of Providence.   Witness:
RICHARD M. FIELD,
Clerk of said Board.

Then, as now, arson was a serious crime which aroused strong public reaction. This 1834 poster offered a reward for the person or persons who attempted to burn the church of the Pacific Congregational Society at Pine and Richmond streets. The revised penal code of 1838 stated that "every person who shall be convicted of arson, shall suffer death, or be imprisoned for life, or for any term not less than ten years." "Malicious burning" carried a sentence of up to ten years, as did "burning to defraud an insurer." Broadside, 1834, courtesy of Rhode Island Historical Society, Rhi x3 4964.

This is a typical "gooseneck" side-stroke engine similar to those used by many Providence engine companies in the 1830s and 1840s. Here a fireward with hat and trumpet shouts commands while six to eight enginemen per side work the brakes up and down to force water through the rotating gooseneck nozzle on top of the air-chamber case and into the hose. When the fire was far from the water source, engine relays were set up to pump water in stages to the piece of apparatus first at the scene of the blaze. The gooseneck, developed in New York, employed the same piston type of pumping that the Newsham engines had used, but it was larger in every respect, having a 6½-inch cylinder and a 9-inch stroke compared with the 4½-inch cylinder and 8-inch stroke of the largest Newsham pump. The gooseneck was also much heavier and bulkier, and so it required twice as many men at the drag ropes and brakes. Painting from Firemen of Industry, courtesy of National Board of Underwriters.

company equipped with the new hose. (A "wash out" occurred when one engine supplied a second with water faster than the second could pump it, causing an overflow; the flooded engine was said to be "washed." Conversely, when one engine could not supply water as fast as it was being pumped out by a second, the first engine was said to be "sucked.") The humiliated Pioneers, who had won many regional firefighting exhibitions, wrote to the Board of Firewards defiantly declaring that "they would not receive water from any machine, nor supply any that use the New York hose." Such obstinacy prompted their dismissal from city service, with their machine and engine-house given to a new company organized under the name of Whatcheer Fire Company No. 8.

However, the Pioneers trumped the firewards by continuing to operate as an independent company separate and apart from the city fire department under their corporate charter of 1836. After an 1847 request for re-admission was denied, they maintained their status until mid-1851, relocating to 166-68 South Main Street, buying an engine, and raising money to support their activities by balls, sales, and private donations. When the Pioneers were received back into the department as Engine No. 11, the fledgling Whatcheer Company was disbanded.

In January 1838 Providence Fire Engine Company No. 9 (Gaspee Company) was incorporated, having commenced duty in 1837 on Carpenter Street in the Smith Hill area, then the extreme northwest section of Providence. Eventually the Gaspee station moved to nearby Pallas Street. This unit originally manned a secondhand Smith engine, but in 1849 it acquired a machine from William Jeffers of Pawtucket, who was beginning his career as one of the nation's most highly regarded fire engine manufacturers. This piece of apparatus later became famous for its appearances in regional firefighting musters and exhibitions. It disappeared mysteriously in 1956 after a fire prevention parade in Providence, and the whereabouts of the valuable and legendary "Gaspee No. 9" is still unknown.

The final engine company created during the volunteer era was Engine Company No. 10. This unit was chartered in 1847 as the Hope Engine Company but quickly assumed the more watery name Atlantic. Located on Knight Street under the captaincy of Thomas Battey

(for whom a nearby street is named), Engine 10 serviced Federal Hill.

The department records for this period show the purchase of two small fireboats for Engine 4 of Fox Point and the strategic placing of two rotary pumps as part of the fire service. One unit was housed at the Steam Mill at Eddy's Point near the foot of Ship Street and another at Howell's Mill at present-day Moshassuck Square.

Another hydraulic innovation occurred in 1836 when the fire service installed its first hydrant (nineteen years after New York City pioneered the technique). Compared with today's high-pressure system, it was a primitive device. In the early nineteeth century water was transported to many West Side households from privately owned fountains via hollowed-out pitch-pine logs with three- and four-inch bores. The Board of Firewards got permission from Field's Fountain Company and other private suppliers to tap into their underground systems of water transport. At various intervals the log pipes were uncovered, holes drilled, and a plug inserted. When blazes occurred, these "fireplugs" could be quickly removed and water pumped from the log to the flames. Wooden pipes gave way to iron in the decade of the 1850s.

An administrative change (overlooked by all historians) also occurred in the 1830s. In a revised set of rules and regulations promulgated in June 1838, the Board of Firewards established the positions of chief engineer and assistant engineer "to direct the department and apparatus at fires." The board then filled these posts from its membership, choosing Henry G. Mumford as first chief. Mumford resigned the top spot a month later to become one of the two assistant engineers. Smith Bosworth succeeded him, followed by Allen Peck (1839-1840). Then, in June 1840, Mumford returned to head the fire service until the board temporarily abandoned this administrative experiment in June 1841.

Providence firemen were actively involved in the politial reform movement that led to the famous Dorr Rebellion of 1842 and the drafting of the present state constitution. As the fire service was dramatically enlarged during the 1820s and 1830s, many landless working-class citizens had become enginemen. Such men were also voteless, since, after 1830, Rhode Island was alone among the states in continuing to require a general real estate

Since 1772, private water companies furnished
a good supply of fresh water to the residents on
the west side of the river. At the time of the city's
incorporation in 1832, four such firms were in
operation—Field's, Rawson's, Dyer's, and
Carpenter's. The oldest (1772) and largest was
Field's Fountain Company on West Clifford
Street (shown here), which then provided thirty-
three thousand gallons per day, mainly for
domestic use. In times of fire emergency, however, these fountains were an important water
source augmenting the wells that had been
drilled for fire purposes after the introduction of
the hydraulion. In 1836 the city's first hydrant
was placed in the system of wooden logs that carried Field's water to Providence homes. Photograph, courtesy of Rhode Island Historical
Society, Rhi x3 4951.

FIRE DEPARTMENT, 1836-7.

RULES AND REGULATIONS

FOR THE GOVERNMENT OF THE

FIRE DEPARTMENT

Of the City of Providence, adopted by the Board of Firewards,

NOVEMBER 15, 1836.

1. THERE shall be elected by this Board, annually, as soon as may be after the election of City Officers in each year, a Chairman, Secretary, Audit, consisting of three members and five directors, to wit: first, second, third, fourth, and fifth Directors.

2. The Chairman shall preside at all meetings of the Board, and may call meetings when he thinks expedient.

3. The Secretary shall keep a fair record of the proceedings of the Board, which shall be read when required; and when requested in writing by the Chairman, or five members, he shall call meetings, and cause the members to be notified thereof, either by personal notice, or by notice in writing left at their places of abode or business.

4. The Audit (of whom two shall be a quorum) shall examine, and if correct, certify all bills of expenses incurred in the Fire Department, under the immediate orders of the Board, or any Committee thereof, and by any Engine Company; provided, that no bill for expense incurred under the order of a committee of the Board, shall be audited, unless certified by such committee; and no bill for expense incurred by any company as aforesaid, unless the same is certified by the Captain, and a fireward thereof.

5. The Directors shall have the control and direction of the Fire Department and Apparatus at fires, and when called out by order of the Board, for exercise; and it shall be their duty to confer together, and recommend general rules for the direction of the Department at fires, which being first approved by this Board, shall be communicated to the respective companies, and published whenever deemed expedient.

6. This Board shall assign the remaining members to such companies and apparatus as they may deem proper, of which the Secretary shall keep a record; and it shall be the duty of any Fireward assigned to any company, to superintend the operations of said company and the apparatus thereof, under the supervision of the Directors aforesaid; and as far as in his power keep the same employed and in operation; and certify all bills thereof which he finds correct and proper.

7. Nine members of the Board shall be a quorum, but a less number may adjourn a meeting.

OFFICERS.

PRESIDENTS OF FIREWARDS.

ELISHA DYER,
RICHARD JACKSON,
CARLO MAURAN,

STEPHEN MARTIN,
BENJAMIN ABORN,
WILBUR KELLY.

Caleb Williams, *Chairman.*
Samuel W Wheeler, *Secretary.*
Benjamin Hoppin,
Benjamin Robinson, } *Audit.*
Samuel W. Wheeler,

Smith Bosworth, 1st,
Earl Carpenter, 2d,
Roger W. Potter, 3d, } *Directors.*
Welcome Congdon, 4th,
Willard W. Fairbanks, 5th,

ENGINES, &c.	CAPTAINS.	FIREWARDS.
Engine No. 1,	Albert Waterman,	John O. Waterman.
Engine No. 2,	James Thurber, Jr.	{ Sylvanus G. Martin, Joseph G. Metcalf.
Engine No. 3,	Henry L. Kendall,	Henry L. Kendall.
Engine No. 4,	James P. Butts,	James P. Butts.
Engine No. 5,	Jonathan S. Angell,	Duty Green.
Engine No. 6,	Joseph W. Taylor,	Hiram Hill.
Engine No. 7,	Pardon S. Pearce,	{ George B. Holmes, Oliver Peirce.
Engine No. 8,	William Aplin,	Asa Pike.
Engine No. 9,	Sylvester Himes,	Edward S. Williams.
Rotary Engine, Steam Mill, Eddy's Point,		Massa Bassett.
Hydraulion No. 1,	James G. Anthony,	{ James G. Anthony, Allen O. Peck.
Hydraulion No. 2,	Allen Baker,	Philip W. Martin.
Stationary Engines in charge of Co. No. 1,	Albert H. Manchester,	{ Benjamin Hoppin, John W. Aborn.
Stationary Engines in charge of Co. No. 2,	William Harris,	{ John H. Greene, Benjamin Robinson.
Hooks and Ladders in charge of Co. No. 1,	Stanton Thurber,	{ Samuel W. Wheeler,
Hooks and Ladders in charge of Co. No. 2.	Sheldon Young,	Robert Manchester.

Attest, SAMUEL W. WHEELER, *Secretary.*

This broadside details the organizational structure of the fire department in the years immediately following Providence's incorporation as a city. The Board of Firewards, which ran the department from 1825 to 1853, was elected in a town meeting by the whole body of freemen until 1832. Under the city charter the board was appointed annually by the City Council, a body that increased its influence over the fire service by creating a Standing Committee on the Fire Department in 1841. Missing from this 1836-37 organizational outline is the building committee, which had been abolished in 1822, and the goods committee, which was discontinued in 1834. Broadside, 1837, courtesy of Rhode Island Historical Society, Rhi x3 4972.

The Providence Washington Insurance Company, founded primarily to furnish marine insurance on the vessels of the bustling port of Providence, evolved from a merger of the Providence Insurance Company (1799) and the Washington Insurance Company (1800). During the long and prosperous presidency of four-term congressman Richard Jackson (1800-1838), this pioneering firm erected offices on a strip of land known as Washington Row, located on the west side of the Providence River (Rhode Island Hospital Trust Bank today occupies this site). In the 1820s and 1830s these offices furnished the setting for many meetings of the Board of Firewards, as the company gradually shifted its underwriting activity from water to fire.

During the presidency of former China merchant Sullivan Dorr (1838-1858), Providence Washington temporarily dropped marine coverage and devoted its energies to the fire insurance field. Dorr (shown here) was the father of the noted constitutional reformer Thomas Wilson Dorr and the husband of Zachariah Allen's sister Lydia.

Today the Providence Washington Insurance Company, the third oldest in the nation, occupies a harmonious, colonial-style building completed in 1949 on a city block bounded by Steeple, North Main, and Canal streets and Washington Place. Here the company maintains a fine collection of fire protection and firefighting memorabilia. Portrait by Charles Loring Elliot, courtesy of Providence Washington Insurance Company, Rhi x3 215.

qualification for balloting and holding office. When traditional efforts to remove this undemocratic restriction were contemptuously rejected by the General Assembly, the disfranchised adult males in 1841 bypassed the state legislature, called their own constitutional convention, and drafted a reformist basic law which they called the People's Constitution. Efforts in 1842 to put this document into effect brought strenuous and resourceful resistance from the incumbent government, controlled by conservatives who styled themselves the Law and Order party.

Thomas Wilson Dorr, a prominent Providence attorney and son of Sullivan Dorr, assumed leadership of the popular movement to modernize Rhode Island's government, and thus the 1842 confrontation between the People's party and the Law and Order party became known as the Dorr Rebellion.

The role of Providence firemen in the initiation of the reform effort was acknowledged by Dorr himself in his brief history of the controversy: "I had no part in originating this last movement," said Dorr. "It was got up, as I have been informed and believe, by the firemen and mechanics of Providence, who deemed themselves as well qualified to vote for their rulers as to do their work and to protect them from conflagration."

The conservatives used their control of the existing government (still operating under the royal charter of 1663) to break up the reform coalition. After the People's Constitution removed the real estate qualification for all adult males, the Law and Order party drafted a constitution that gave native-born citizens broad suffrage but retained the real estate requirement for naturalized citizens—which in 1842 meant Irish Catholics. Then the conservatives urged landless native workmen to reject the People's basic law, alleging it would allow Irish immigrants to take over the government of the state. Many firemen and other working-class rebels were frightened off by this maneuver and abandoned Dorr.

In addition, the conservative state legislature passed a statute in 1842 increasing the authorized strength of the Providence Fire Department to twelve hundred men, about double its existing size. The purpose of this law was also to persuade working-class agitators to abandon the re-

SUFFRAGE MEETING IN THE 6TH WARD!

The Freemen and Non-Freemen of the SIXTH WARD, who are in favor of Abolishing the Charter granted to this State by King Charles the 2d, and adopting a Republican form of Government, are requested to meet at the *WARD ROOM, in Summer Street,* on

Wednesday evening, April 14
at half past 7 o'clock.
☞ **A general and punctual attendance is requested.**

This handbill, circulated by the Rhode Island Suffrage Association in 1841, invited citizens to a mass meeting in Providence's old Sixth Ward (the present Hoyle Square section) to protest the continuance of the real estate qualification for voting, a device that then disfranchised nearly 70 percent of Providence's free adult males, including many members of the enlarged fire department. This agitation culminated in the Dorr Rebellion, America's most famous democratic uprising. Providence firemen played a major role in this reform movement. According to one contemporary account, the Suffrage Association was formed by those voteless men "who did military duty and worked the fire engines." Broadside, circa 1841, courtesy of Rhode Island Historical Society, Rhi x3 4993.

form cause. Since the formation of the first engine companies in 1763, one of the major advantages of the fire service was that a substantial number of firefighters were exempted from militia and jury duty. From the mid-1820s onward all special acts creating new fire companies prominently stated this privilege. When the duties of the state militia were dramatically increased in January 1840, alternative public service in the fire department became even more attractive. Some historians claim that the stringent militia law of 1840 was the spark that ignited the working-class movement leading to the Dorr Rebellion. Thus in January 1842 the General Assembly appeased many discontented Providence citizens by allowing them to enter the fire service. Then, in 1844, with the rebellion over, all fire companies in Providence received the militia and jury duty exemption for every one of their members, a concession designed in part to forestall further political agitation.

Despite the concessions made to volunteer firemen during and after the Dorr Rebellion, their greatly enlarged number, coupled with the extension of suffrage, made them a political force to be reckoned with by the City Council. In June 1841 that body had made an attempt at gaining greater supervision over the volunteers by establishing a Standing Committee on the Fire Department, consisting of three members of the Common Council and one member of the Board of Aldermen.

During the later 1840s and early 1850s—turbulent times in the emerging cities of America—the volunteers got increasingly unruly and successfully resisted periodic attempts by the council to control them. In the aftermath of the hose dispute, the Pioneers battled the city government to a standoff from 1846 to 1851, during which period they functioned as an independent company. In August 1852, firemen quickly forced the repeal of a March ordinance giving the council greater financial supervision over the department and banning several volunteer practices that violated good order. That October, troublesome Engine Company No. 5 was disbanded by the firewards "for refusing to do duty," and in 1853 Eagle Company No. 1 of Olneyville had its engine confiscated "in consequence of disturbances there at fires."

The veteran Zachariah Allen perceptively summed up the problems besetting the department in an 1854 essay that merits extensive quotation:

> The abuses complained of have originated from the disproportionate number of the youthful and imprudent members of the fire-engine companies who have taken the places deserted by the older, wealthier, and more sedate citizens.... Disinclined to social hilarity, to arraying themselves in uniform, and to joyous excursions abroad, and annual meetings at home, the older members have left the management of the operations for extinguishing fires in the hands of the youthful and imprudent. Under these circumstances, it is not a matter of surprise that frequent excitements and excesses should have occurred, origi-

The most celebrated house fire in early Providence history occurred on November 20, 1849, at the Benefit Street mansion of widow Anna Jenkins, the only granddaughter of Moses Brown. The Jenkins house, built in the 1780s by merchant John Innes Clark, had been visited by such notables as George Washington. The blaze that leveled the historic structure killed Mrs. Jenkins (aged fifty-nine) and her eldest daughter Sarah (aged twenty-two). Survivors included daughter Anna, son Moses, and several servants, who were awakened by the family dog.

This faithful animal was memorialized by the first bronze statue cast by the Gorham Manufacturing Company. Called "The Sentinel," the statue, executed by Thomas Frederick Hoppin, originally stood on the front lawn of the Thomas Hoppin House (young Anna married Thomas Hoppin and they built their mansion, still standing at John and Benefit streets, on the site of the burned structure). Today Providence's first fire dog—which was displayed at the Crystal Palace in London and won a gold medal from the New York Academy of Design—stands in the Roger Williams Park Zoo, where it furnishes entertainment for young children such as the one shown here. Photograph, 1939, courtesy of Mary and Julia Conley.

nating from the maddening impulse of stimulating drinks which have been injudiciously distributed by those having property endangered near the scenes of conflagration.

The time has now arrived when the reorganization of a new system must be forthwith adopted. It has become an absolute necessity that the most respectable and wealthy citizens must again enroll themselves as formerly, and take into their own hands the charge of protecting their own property. Or if they choose to continue to remain quietly in their beds, they must pay an adequate number of men for their services to work for them. . . .

It must be far pleasanter to every generous mind, to pay an equivalent in taxes, for the services of the workingmen of the city in extinguishing fires, than to feel a weight of obligation for an undefined debt of gratitude. Embarrassed by this sense of gratitude for unrequited services of the firemen, the City Council have made numerous grants of large sums for their gratification, amounting to nearly fifty thousand dollars in two years, in the purchase of new and fanciful fire-engines, new and spacious halls, resembling European club-houses, decorated with curtains, mirrors, chandeliers, gildings and paintings etc. These appropriations have been profitlessly—not to say demoralizingly and perniciously—wasted for the purpose of sustaining the unpaid system, which has consequently been a costly one.

Where there is no reward for obeying orders, and no penalty for disobeying them, there can be no regular discipline in the organization of the members of the Fire Department, and no certain reliance on their cooperation. They individually are independent of control, and feel themselves to be at liberty to stop to dispute about precedency of position, or to fight whilst a conflagration is raging.

The old system being actually disorganized, it now remains to carry into effect, energetically, the system of a Paid Fire Department.

In June 1853 the General Assembly, moved by sentiments similar to those expressed by Allen, passed an enabling act conferring upon the Providence City Council broad powers to legislate in fire department affairs. One month later the council passed an ordinance replacing the firewards by a chief engineer and six assistants. These seven council appointees constituted a Board of Engineers to govern the fire service with the advice of the Standing Committee on the Fire Department. The law

Water Witch Engine Company No. 6 was incorporated by the General Assembly in 1834. Within a short time "the Sixes," as they were called, developed a reputation second only to the volunteers at Hydraulion No. 1. The Water Witch Company began with a membership of forty-five. Included on its initial roster was Joseph W. Taylor, later to become the department's chief engineer.

Shortly after it incorporated, the volunteer company received its side-stroke hand pumper. Constructed by James Smith of Waterford, New York, the Water Witch engine was elaborately decorated and topped by a banner that boasted its motto—"Actuated by Benevolence, Impelled by Emulation." The Sixes developed a reputation for military-style discipline and were the first of the volunteer companies in Providence to adopt uniforms. The spectacle of the Water Witch racing down Benefit Street headed by her smartly dressed company in blue trousers and red flannel shirts inspired local composer James M. Bradford to immortalize their exploits in song (upper right).

The company constructed a station near the corner of Benefit and College streets in 1852 (right). However, the grand era of the volunteer companies was about to close. In 1863 the Water Witch enginehouse was razed. Three years later the venerable side pumper was sold to the A. W. Sprague textile firm in Cranston and added to its fire company. Later returned to a group of former Water Witch Company members, the engine was presented by them to the Providence Veteran Firemen's Association. Sheet music, 1853, by James Bradford, courtesy of Rhode Island Historical Society, Rhi x3 3103. Photograph, 1860, courtesy of Rhode Island Historical Society, Rhi x3 4499.

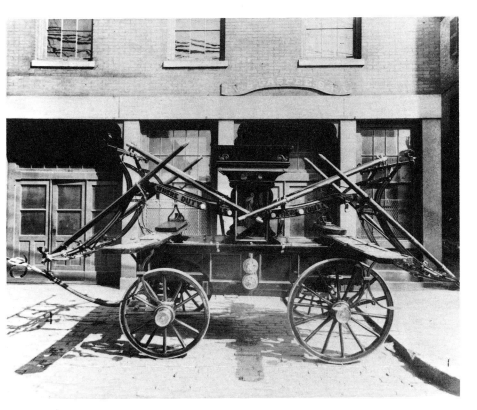

The Gaspee hand engine was typical of the standard piece of firefighting equipment just before the introduction of the steam fire engine. Built by William Jeffers of nearby Pawtucket in 1849, the Gaspee was an end-stroke "double deck" engine that required the work of as many as twenty men. Folding upper platforms in front and back held five men each. When the Gaspee arrived at a fire, the break arms were swung out and hoses attached to the suction and discharge valves.

After seeing service in Providence for ten years, in 1860 the Gaspee was sold by the city and went to New London, Connecticut. During Rhode Island's celebration of the 250th anniversary of its founding, a group of survivors of the Gaspee Volunteer Company initiated a search for the old engine. Discovering it in Putnam, Connecticut, the group arranged for its purchase. It became one of the most valued artifacts of the Providence Veteran Firemen's Association. The photograph shows the Gaspee in front of the association's headquarters on South Main Street. *Photograph, circa 1906, courtesy of Rhode Island Historical Society, Rhi x3 4451.*

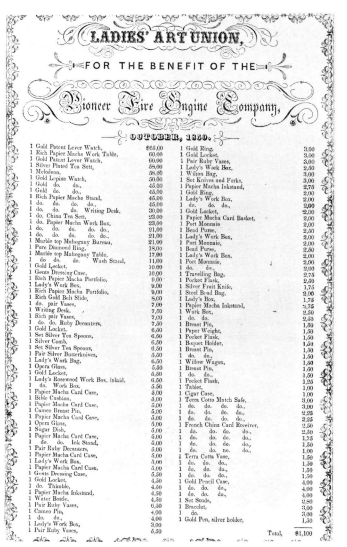

Some volunteer companies raised money for their activities by firemen's balls, fairs, auctions, and sales such as the one noted on this broadside. The recipient of the generosity of the Ladies' Art Union was the Pioneer Fire Engine Company, the city's most cantankerous and controversial unit. The Pioneers were formed in 1836 as Engine No. 8, housed on Benefit Street at Transit, and during their first decade of operation both their service record and their performance in regional firefighting competitions were exemplary.

In 1846, however, the Pioneers entered into a heated dispute with the firewards, who were slow in furnishing them with a new brand of improved high-caliber hose. When the controversy resulted in the dismissal of the company from the department, the Pioneers decided to perform as an independent unit. For the next five years they used several fund-raisers, such as this sale, to buy their own engine, hose, equipment, and firehouse on South Main Street. In 1851 the schism was healed, and the Pioneers were reinstated by the Board of Firewards as Engine No. 11. *Broadside, 1850, courtesy of Rhode Island Historical Society, Rhi x3 4973.*

gave the board power to disband recalcitrant engine companies, take possession of any city property held by such companies, and hire men to perform fire duties formerly furnished by the dissolved unit. The council chose as chief engineer former first fireward Joseph W. Taylor, who had begun his departmental duties in 1834 as captain of Water Witch Company (Engine 6).

In the immediate aftermath of this limited reorganization, the volunteers engaged in a riot that hastened their own demise. Some knowledge of the social climate of the early 1850s is necessary for an understanding of this disruptive episode. Several strong emotional currents were then sweeping the North, especially the campaigns against slavery, Irish Catholic immigrants, and "Demon Rum." Two of these highly charged issues—nativism and temperance—contributed to the disturbance that sounded the death knell for the volunteers.

The impact of drink upon the conduct and efficiency of the city's firefighters was alluded to by Zachariah Allen in the essay quoted above. In Providence (and elsewhere) a custom had developed whereby shopkeepers and saloonkeepers in the vicinity of a fire would fortify their protectors with large quantities of free liquor. These spirits intensified the already keen competition and rivalry amongst the several engine companies. When their activities conflicted, fights erupted and the fire was ignored. This situation occurred on the morning of October 10, 1853, when the volunteers sped to douse a blaze in the Arnold Block on North Main Street. First the Pioneers and the men of Water Witch went at it; then the Niagara Company, for reasons unclear, descended upon a hapless Irishman named Neal Dougherty, a paid teamster (but not a member) of Gaspee No. 9. Despite the attempted intercession of Chief Engineer Taylor, Dougherty was beaten, pelted with stones, chased, and beaten again during a desperate flight from the foot of Waterman Street to the old Union Railroad Depot on Exchange Place, where he was clubbed with an iron bar. The savagely assaulted man died in a doctor's office within the hour.

A newspaper report of the coroner's inquest stated that the attack was "not provoked by the victim, but was the brutal act of men infuriated by liquor." Three Niagara men went to trial, but contradictory testimony of witnesses about the melee prevented their conviction.

The victim of this atrocity was an Irish Catholic immigrant, a fact which undoubtedly contributed to the vehemence of the attack. For nearly two decades, as Irish poured into the cities of the Northeast, bloody clashes had occurred between these newcomers and fire companies composed of WASP natives drawn increasingly from the working class. In Philadelphia, Boston, New York, Baltimore, Cincinnati, and elsewhere, Irish gangs and engine companies had done battle, with dozens of deaths and hundreds of serious injuries on both sides. In the infamous Philadelphia riot of 1844, firemen burned "one nunnery, one school-house and three large Catholic churches," according to one newspaper account. Providence's Dougherty incident, though brutal, was minor compared to the disruptions in other major cities, and it did not occur until the anti-Catholic crusade known as Know-Nothingism had nearly reached its peak.

The city fathers and other sober, responsible citizens were outraged by the slaying, and local public opinion turned increasingly against the volunteer system and towards a paid force. Steps in that direction had in fact already been taken during the five years preceding the 1853 riot. In 1848 the Board of Firewards had voted to pay an annual stipend (averaging sixty dollars) to the stewards (record keepers) of the engine companies and hired a superintendent of repairs (Allen Baker) at an annual salary of eight hundred dollars to maintain the department's property and apparatus. In the following year the board decided to pay a cash stipend (in lieu of the traditional practice of furnishing refreshments) to those companies engaged at a fire of more than one hour's duration, and to prevent the dissolution of Hook and Ladder Company No. 2, it authorized a fifteen dollar-per-year salary for each of the unit's fifteen members.

A council committee created in the aftermath of the riot compiled a report scathingly critical of the status quo. It urged Providence to follow the example of Cincinnati, which had established a paid department on April 1, 1853. The City Council, which had been called into existence after the riot of 1831, responded with an ordinance,

This mid-nineteenth-century lithograph epitomizes the volunteers in action. To the extreme left is the torchbearer (a position held by such Providence department luminaries as Thomas Aldrich), followed by the fireward, trumpet in hand. At the dragropes of the side-stroke gooseneck hand engine are the diverse members of the firefighting corps, from top-hatted businessman to blue-collar worker. Lithograph by Thomas and Wylie, courtesy of the Library of Congress.

effective March 1, 1854, creating a paid fire service and making Providence the second major city in the nation to adopt this innovation.

The volunteer era was over. Though its end was ignominious, sentimental memories engendered by years of service and camaraderie could not be effaced by the old department's demise. The Water Witch Company soon made a decision to preserve its heritage by establishing a small firefighting museum. On August 24, 1854, the *Providence Journal* described that decision with a touch of pathos:

> The members of Water Witch Engine Company No. 6, upon the breaking up of the volunteer fire department, found their associations too agreeable to be surrendered so readily, and they accord-ingly determined to remain together. . . .
> They have taken an elegant suite of rooms in Arnold's Block, and have fitted them in a style of great splendor, with frescoed walls, damask curtains, rich carpets, furniture and chandeliers. On one side, protected by glass cases, are numerous trophies and memorials connected with the history of the company, some of the ornaments of their machine, and the trumpets of their favorite officers—all arranged with much taste and present a beautiful show.

After a full century of always exciting and often heroic exploits, the good memories died hard.

In the early 1850s the fire department went through a series of alterations. This March 1852 ordinance vainly tried to bring the volunteers under more effective control because the corporate charters several companies had received from 1834 onward made them semiautonomous and difficult to regulate effectively. Under this attempted reorganization, seven firewards appointed by the council and designated by numerical rank were to constitute the governing board of the department. The "first fireward" was, in effect, the chief. Because of recent extravagances by several companies, the ordinance provided for close council supervision of departmental finances.

Reflecting a growing rowdyism by enginemen, the new law banned cards, other gaming implements, and "spirituous liquors" from the enginehouses, provided fines for "breach of the peace or other violation of good order," forbade any company to "beat its gong, ring its bell, or use any musical instrument, while returning from a fire," and banned driving on sidewalks. With the Pioneers in mind, the ordinance also ended the "practice of permitting volunteer associations, not belonging to the Fire Department, to mingle in the duties thereof."

When the volunteers vehemently protested against the law and flexed their political muscle, the council backed down and repealed it in August 1852. After the fire riot of October 1853, however, public sentiment turned against the volunteers, and a comprehensive ordinance establishing a paid department passed the council in 1854. Broadside, 1852, courtesy of Rhode Island Historical Society, Rhi x3 4961.

AN ORDINANCE
ESTABLISHING THE
FIRE DEPARTMENT.

It is ordained by the City Council of the City of Providence. as follows, viz:

Section 1. The Fire Department of the City shall consist of seven Fire-wards and as many Hose, Hook and Ladder and Engine Companies as the City Council may, from time to time elect, so that the number of Firemen (twelve hundred) prescribed by law, be not exceeded by the whole Fire Department. The seven Fire-wards shall be designated as First, Second, Third, Fourth, Fifth, Sixth and Seventh Fire-wards, and take rank accordingly—the first holding rank by election, the others determining their rank among themselves. They shall be elected by the City Council ; the first, from the city at large ; the other six, one from each ward ; and from the number of Fire-wards so elected, the City Council shall elect three Presidents of Fire-wards. The City Council shall also elect the members of the different companies ; but each company shall elect its own officers.

The whole number of Fire-wards shall constitute a Board for the purpose of making regulations for the government of the Fire Department, trying cases of insubordination or violation of said regulations, and transacting such other business belonging to the Department, as may come before them.

Sec. 2. Each company shall be commanded by a foreman and such other officers as they may choose, who shall have special charge of the companies, engine-houses and apparatus under their command ; report any deficiency, defect or want of repairs, to the First Fire-ward ; see that all the regulations of the Department relating thereto, are strictly observed ; and cause ell orders emanating from the Firewards, directed to their companies, to be carried into immediate execution.

The whole Fire Department shall be under the command of the First Fire-ward ; or, in his absence or inability to act, of the highest Fire-ward present. It shall be the duty of the Fire-wards to repair immediately to all fires within the limits of the city ; and their orders shall be promptly obeyed in all matters pertaining to the preservation of property, to the location and use of engines, to the extinguishing of fires, and in all other things belonging to the Fire Department. All disputes and controversies that may arise in matters relating to the Fire Department between any of its officers, members or companies, shall be summarily decided by the Fireward in command, whose decision, shall be, for the time, obeyed ; but if any party be aggrieved or injured by such decision, he or they may appeal to the Board of Firewards whose decision shall be final.

Sec. 3. The First Fireward shall have the general supervision of all repairs ; he shall see that the Engine-houses and the Reservoirs of water for extinguishing fires ; the Engines (stationary or otherwise,) the Hose, Hooks and Ladders, and their carts or trucks, and all the fire apparatus, are kept in good order and ready for immediate use. He shall make himself familiar with the location of the Reservoirs, the depth of water in them, and the length of Hose necessary to reach it ; and furnish a list of the same, with this information annexed, to the foreman of each company ; and he shall make arrangements to secure the ringing of bells in different sections of the city, in alarms of fire ; and report monthly to the City Council, the amount of all the expenditures of the Fire Department. Any application to the Board of Fire-wards for the purpose of building Engine-houses, or Reservoirs, or buying lots for the same, or purchasing Engines, shall be considered by them and reported to the City Council. All bills against the Fire Department shall be examined and certified by an Audit appointed by the Board of Fire-wards, before being audited by the City Auditor. The Board of Fire-wards shall annually report to the City Council, the state of the Fire Department ; with a list of fires in the city in the year just passed, and such statistics in relation to losses, insurance and causes of fire, as they may be able to procure ; and suggest such alterations and improvements as they may deem advisable.

Sec. 4. It shall be the duty of the whole Fire Department, unless specially excused, to proceed, as soon as practicable, with their Engines and apparatus, to every fire that breaks out in the city. No company shall take its Engine out of the city without the consent of the First Fireward. Not more than two companies, with their engines, shall be absent from the city at any one time on a pleasure excursion ; nor shall more than one third of the whole number of Engines leave the city at any one time, on any pretence whatever. No company shall beat its gong, ring its bell, or use any musical instrument, while returning from a fire. In case of any breach of the peace or other violation of good order, on the part of any fireman, while on duty, it shall be the duty of the officer in command at the time, to report the name or names of the person or persons so offending, to the Board of Fire-wards ; who shall inquire into the matter, and shall have power to expel the offender or offenders from the Fire Department.

Sec. 5. No cards or other gaming implements, shall be used in any Engine-house ; nor shall any disorderly conduct be permitted, nor spirituous liquors be used therein ; and no Engine-house shall be opened on Sunday, except in case of alarm of fire.

Sec. 6. In case any Foreman, Assistant Foreman, or any other Fireman, having command of any company, when on duty, shall disobey or refuse to obey any order or direction given by any Fire-ward, he shall for such offence, be suspended or expelled from the Fire Department, at the discretion of the Board of Fire-wards.

Sec. 7. If any company shall become inefficient for want of numbers, it shall be the duty of the Board of Fire-wards to notify the Foreman of such company, that, unless within a reasonable time they recruit their company to efficient numbers, they will be disbanded and their Engine taken from them ; and if such company shall fail so to recruit their numbers as to answer the requirement of the Fire Department, the Board of Fire-wards shall disband the company and take possession of the Engine.

Sec. 8. Whereas, the practice of permitting volunteer associations, not belonging to the Fire Department, to mingle in the duties thereof, has been found pernicious in its results, it shall be the duty of the Fire-ward in command, to order from the ground, at fires, all such companies.

Sec. 9. No Fire Engine or Hook and Ladder Truck, or Hose Cart shall, in going to, or returning from any fire, or at any other time, be run, driven, wheeled, drawn or placed upon any side-walk, except by the special order of a Fireward, under the penalty of Ten Dollars for each offence, to be forfeited and paid by every person aiding or assisting in the violation of this Ordinance.

Sec. 10. No person shall be eligible to the office of Fireward, after he shall have arrived at the age of sixty years.

Sec. 11. This Ordinance shall go into effect on the first Monday in June, 1852 ; at which time all Ordinances, Resolutions and Votes of the City or Town of Providence, in relation to the Fire Department, shall be repealed.

Passed March 15, 1852.

A true copy—witness,

ALBERT PABODIE, City Clerk.

CITY ORDINANCE.

An Ordinance in amendment of an Ordinance entitled " An Ordinance establishing the Fire Department."

It is ordained by the City Council of the city of Providence as follows :

Section 1. Each Company may recommend its own Foreman, who, when elected by the City Council, shall be duly sworn and commissioned to the faithful discharge of his duties.

Passed March 29, 1852.

A true copy—witness,

ALBERT PABODIE, City Clerk.

Born in Little Compton, Joseph W. Taylor moved with his family to Providence when he was ten years old. He entered the fire service in 1834 as a charter member of the Water Witch Volunteer Company. Five years later Taylor became a fireward. He remained in that position until 1852, when the council chose him president (i.e., presiding officer) of the Board of Firewards. In that balloting he bested Thomas Aldrich, another veteran fireward (and future chief), by a vote of 18 to 10.

In July 1853 the Board of Firewards was abolished. Its duties were assumed by the newly created offices of chief engineer and six assistants. Taylor's position as board president and his ten-year service as board secretary (1838-1848) made him the logical choice as chief. Three months after Taylor's appointment an altercation between the Niagara and the Gaspee volunteer fire companies at the Arnold Block fire intensified demands to institute a paid department. On January 25, 1854, the City Council adopted a resolution abolishing the volunteer system, and on March 1 a new ordinance went into effect establishing a paid force composed of permanent and call firemen not to exceed 450 in number. Just before leaving office in 1859, Taylor contracted to purchase two steam fire engines, opening a new era for the department. Chief Taylor, well liked by the rank and file, died on September 20, 1876. Engraving by Ryder-Dearth, courtesy of Rhode Island Historical Society, Rhi x3 4969.

Despite the circumstances surrounding its demise, the volunteer era of the Providence Fire Department was a colorful and exciting phase of local fire history. The department's first historian, Governor Elisha Dyer, surveyed this period in an address delivered to the quarterly meeting of the Veteran Firemen's Association in October 1885 and eventually published as a pamphlet entitled Sketch and Reminiscences of the Providence Fire Department from 1815 to 1854. Dyer was well versed in his subject. His father Elisha (the elder) had worked with Zachariah Allen in 1822 to procure the first hydraulion, and both Dyers were long active in the volunteer fire service. Historian Dyer, shown here, served as governor of Rhode Island from 1857 to 1859. His own son and namesake was governor from 1897 to 1899 and also served as mayor of Providence in 1906.

A study of the rosters of the firewards and the various volunteer companies shows that at least ten members of the volunteer system served as governors of Rhode Island—Stephen Hopkins, Nicholas Cooke, William Jones, James Fenner, Henry Bowen Anthony, William W. Hoppin, William Sprague, Elisha Dyer, Jr., James Y. Smith, and Seth Padelford. Photograph, 1868, courtesy of Rhode Island Historical Society, Rhi x3 4966.

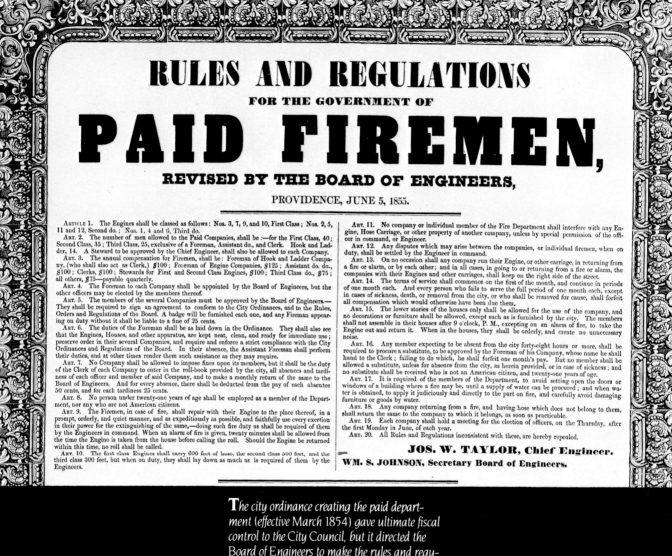

The city ordinance creating the paid department (effective March 1854) gave ultimate fiscal control to the City Council, but it directed the Board of Engineers to make the rules and regulations governing the fire service. This broadside of twenty articles was the first directive detailing the duties of firemen under the new system. All firemen and their foremen were to be board appointees. The rules also addressed several abuses by the volunteers. Coming at the height of the Know-Nothing movement, the regulations specifically restricted the paid service to Ameri-

chapter TWO

The Age of the Steamers, 1854-1899

The Providence Fire Department became a paid organization effective March 1, 1854. The council ordinance establishing the new department retained the post of chief engineer, and Joseph Taylor continued to fill it. The law also provided for five assistant engineers, and these, with the chief, constituted a new Board of Engineers to supervise and regulate the department. Because of the recent extravagances of the volunteers, financial control rested with the council. The board was merely an advisory body in the area of departmental expenditures, having final say only in the sale or disposal of unserviceable fire apparatus.

The most notable change in the department was its size. During the final years of the volunteer regime, the number of firemen exceeded 1,100, just below the statutory limit. The 1854 act authorized a force of not more than 450. To meet this restriction, five companies were scaled down or eliminated. When the streamlining was complete, eleven hand-engine companies and one hook-and-ladder unit remained, with a total complement of 432 men.

Despite the change there was a considerable degree of continuity in the department, because most of the new paid men were drawn from the ranks of the more able and efficient volunteers. In addition, most of the old engine companies maintained their identity, their station, and their apparatus. The major exceptions to this rule were the two hydraulion units. Hydraulion No. 1 became Engine No. 1 because of the recent dissolution of Olneyville's Eagle Company, and Hydraulion No. 2 became Columbia Hand Engine Company No. 12, located on Haymarket Street. The old hydraulion engines were both disposed of to buyers unknown. There was no Engine 8 because of the Whatcheer unit's demise in 1851.

The new ordinance provided each company with a foreman appointed by the Board of Engineers, but it allowed company members to elect their other officers. In 1854 the foremen received an annual salary of $125; the assistant foremen, $100; the stewards and clerks of companies, $100; and the enginemen, $75. Finally, the statute divided Providence into fire districts corresponding to the existing police or "watch" districts.

In the first two annual reports of the paid department (June 1854 and 1855), Chief Taylor praised the new system, observing that "contrary to the general expectation that paid firemen would be inactive and uninter-

ested, they have manifested from the first an activity and interest which has never been excelled." Their "decorum," order, "harmony," and conduct, he exulted, were "excellent." To prove his point concerning the superiority of the paid force, Taylor prepared a table of fire statistics for the last fifteen months of the volunteer era and the first fifteen months of the new regime. His figures showed a decrease of nearly 400 percent in the value of real and personal property lost by fire.

The chief's only laments in these glowing reports were about the archaic nature of the alarm system and the lack of a sufficient water supply. To alleviate the latter defect, the department had installed four new stationary engines (bringing the number to six) and "about 2,000 feet of iron pipe running from them with hydrants connected." By the end of the decade the city had laid 4,780 feet of pipe connected to twenty-six hydrants drawing water from forty-eight "reservoirs" and six wells. The first hose company, established in 1856, was disbanded by the board after a year's trial.

In June 1859 Taylor's final report indicated that he had followed the advice of the council's Committee on the Fire Department and contracted for two steam engines, one from the Philadelphia firm of Reanie and Neafie, the other from the Silsby Manufacturing Company of Seneca Falls, New York.

It was left to Taylor's successor, Thomas Aldrich, to preside over the actual introduction of the steamers into the fire service in mid-September 1859. Aldrich, like his predecessor, had been a president and a fireward under the volunteer system, but Taylor had edged him out in the 1853 balloting for chief engineer. In June 1859 the persistent Aldrich turned the tables in a close contest, 19 to 15. The new chief, an accountant and cotton broker by trade, had a short but eventful tenure.

Shortly after Aldrich's accession the Silsby engine arrived. Aldrich appointed James Salisbury, Jr., the foreman of the ten-member Steam Engine Company No. 1, and twenty-five-year-old Ira W. Winsor (who would play a prominent role in department affairs in the late nineteenth and early twentieth centuries) was designated "first hoseman" for this new rotary machine. On the heels of the Silsby came a single-acting piston engine from Reanie and Neafie, and Steam Engine Company No. 2, under Henry W. Rodman, was created. Early in 1860 the city purchased another Silsby steamer and formed Steam

Joseph M. Wightman, a former machinist and manufacturer, wrote this February 1859 letter to Providence Mayor William Rodman (1857-1859) to introduce Robert Bickford, a representative of a Boston-based steam-engine company, and to urge consideration of Bickford's machine. On August 31 and September 1, 1858, a "grand-steam engine contest" had been staged on Boston Common, with the Philadelphia firm of Reanie and Neafie besting several Massachusetts steamers. Disregarding Wightman's recommendation, Providence selected engines from Reanie and Neafie and the Silsby Manufacturing Company of Seneca Falls, New York, to launch its experiment with steam in the late summer of 1859. Wightman, who had just finished a term as city alderman, was elected Boston's mayor in December 1860. Letter, February 19, 1859, courtesy of Mildred S. Longo.

Thomas Aldrich developed an early interest in the fire service and first saw action as a torch-bearer for one of Providence's volunteer companies. In 1822 the seventeen-year-old Aldrich joined Hydraulion Company No. 1. Twelve years later he was elected the unit's second lieutenant. In 1837 the City Council appointed him a fireward, a post he held until 1851, often doubling as department auditor. During the last years of his tenure, Aldrich chaired the Board of Firewards.

This ambitious Providence native was also an astute businessman who teamed with Peleg Gardiner to form a cotton-brokerage firm. In June 1859 Aldrich was called to duty again, this time as chief engineer of the department. Although holding this position for only three years, he introduced the steam fire engine to Providence and oversaw the installation in 1860 of the city's first telegraph fire alarm system. Aldrich returned to the cotton business following his tenure as chief. He died of a heart attack on February 16, 1883, at the age of seventy-seven. Engraved portrait by Ryder-Dearth, courtesy of Thomas Aldrich.

This rare photograph shows one of the first three steam fire engines purchased by the city of Providence. This rotary steamer, built by the Silsby Company of Seneca Falls, New York, arrived in Providence early in 1860 and was assigned to Steam Engine Company No. 3 on Summer Street in South Providence under the direction of Foreman (later Chief) Oliver E. Greene. Photograph, courtesy of Rhode Island Historical Society, Rhi x3 4927.

Fire King Steam Engine Company No. 3, created in 1860, was the successor to volunteer Hand Engine Company No. 5, whose motto—"Mid the Raging Flames the Fire King Reigns"—was adopted by the newly organized steam-engine company, shown here. Their station, built on Summer Street in 1840, remained Fire King's headquarters until 1875, when the company moved to a more modern facility on nearby Pond Street. In that year the unit was transformed into Hose Company No. 3 because the recent introduction of an abundant water supply had made an increase in hose companies advantageous. Photograph, 1860, courtesy of Rhode Island Historical Society, Rhi x3 4957.

By the early 1860s the Providence department boasted both rotary and piston-driven steam fire engines. The piston pumpers (like the one shown (right) generated steam (S) that entered the steam cylinder, forcing the piston (P) downward into the water cylinder. This piston drew water through the intake (W) and forced it up into the air chamber (A). The air, becoming greatly compressed against the dome by the rising water, exerted a constant pressure and sent a steady flow through the line of hose (H).

The rotary engines were driven by two horizontal cams that were turned by steam pressure, drawing and then forcing water through connecting hoses. Diagram and text from John V. Morris, Fires and Firefighters.

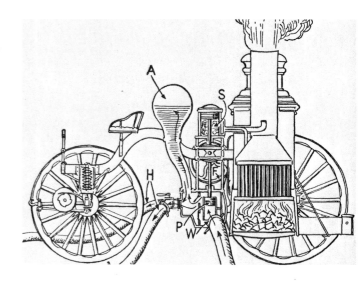

Engine Company No. 3, directed by Oliver E. Greene, a future chief.

These new thirty-five hundred dollar machines, each manned by ten men, caused the disbandment of four hand-engine companies, Nos. 1, 5, 11, and 12, each with a complement of over thirty men—a total reduction of nearly one hundred firefighters from the paid force. The weight of these early steamers—about seven thousand pounds—necessitated the hiring of horses and drivers to pull them.

As new steamers were introduced during the decade of the 1860s, hand engines were discarded and the ranks of the department were thinned still further. In 1867 the last four hand pumpers (three manufactured by New York's Button and Company and one made by Pawtucket's William Jeffers) were decommissioned and consigned to musters and exhibitions. An all-steam fire service had progressed from experiment to reality.

In addition to the introduction of steamers, the Aldrich regime was notable for the establishment of a modern telegraph alarm system in the summer of 1860. Adapted by Providence councilman and fireman Charles E. Carpenter from a system developed in Boston by William F. Channing and Moses G. Farmer, this fire alarm was hailed by Aldrich in his 1861 report for "doing all that was promised or expected by its originators—giving rapid information to firemen whenever their services are required."

In June 1862 thirty-three-year-old Charles H. Dunham became chief engineer, a post he held with competence for three years. During Dunham's tenure the department organized a second hook-and-ladder company and equipped it with a truck built by Providence's own Moulton and Remington, a firm which would supply the department with most of its hose carts and hook-and-ladder trucks for the next quarter century. In 1864 the Armory Hose Company, a volunteer group composed of em-

ployees of the Providence Tool Company, was formed "to render service whenever a fire occurs near their premises" on Wickenden Street. Also in 1864, at Dunham's suggestion, the city took steps to purchase, rather than lease, the horses to pull its steamers. A September 12 ordinance provided for the hiring of hostlers (groomers of horses) and assigned one to each steam-engine company. However, the first four horses to pull Steamer No. 7 were not acquired until early in 1867.

On July 10, 1865, Assistant Engineer Dexter Gorton succeeded the ailing Dunham (who died on August 29) as chief. During Gorton's three-year tenure the transition to steam was completed with engine purchases that included machines built by two prominent Pawtucket manufacturers, William Jeffers and Cole and Brothers. In addition, a third hook-and-ladder company was organized, several new stations were constructed, three horses were purchased for each steam engine and one for each hook-and-ladder truck, and several cisterns were constructed. In February 1868 a modest restructuring of the department occurred when the number of assistant engineers was reduced to two. In that year the old (and largely honorific) title of president of the firewards was abolished, with Charles E. Carpenter, Sturgis P. Carpenter, and Dexter Gorton being the last to hold that designation.

Much of the technical and physical growth of the department that occurred under Gorton and his successor, Chief Oliver E. Greene, resulted from the energetic efforts of Thomas A. Doyle, a man regarded by historians as the city's most productive mayor. An independent-minded Republican of Irish Protestant stock, Doyle served from June 1864 to June 1869, from June 1870 to January 1881, and from January 1884 until his death in office on June 9, 1886—a total of eighteen years as chief executive. Among Doyle's many accomplishments were the commencement of a sewer network (1874) and the construction of City Hall (1878). His contributions to the fire

Charles H. Dunham served as chief engineer from June 1862 until July 10, 1865, when sickness forced him to vacate his post. He died seven weeks later at the age of thirty-six. Ironically, Dunham was the youngest chief in the department's history. During his tenure a new station was constructed on Haymarket Street to house a new Silsby rotary steamer and a hook-and-ladder truck that was purchased by the department in 1863. Dunham also expanded the city's network of cisterns, but the days of wells as the primary source of water for firefighting were numbered. Engraved portrait by Ryder-Dearth, courtesy of Rhode Island Historical Society, Rhi x3 4967.

service included the construction of a central station (1874) at the east end of Exchange Place, housing Hose No. 1, Hook and Ladder No. 1, Protective Company No. 1, and the offices of the chief engineer; the building of several two-story brick stations to house the new steam apparatus; and the establishment of a modern water system, a project staunchly advocated by Zachariah Allen and Chief Taylor in the 1850s.

On December 1, 1871, water from the Pawtuxet River pumped by a station in the Pettaconsett section of Cranston reached the taps and the hydrants of Providence. In 1875 a second reservoir was completed on Hope Street on the present high school grounds, and in 1889 a third "high service" reservoir became operative in the Fruit Hill section of North Providence. These abundant sources of water triggered a gradual phaseout of the system of wells and cisterns that had been the department's principal water supply since the mid-1820s. By 1871 these wells numbered sixty-nine.

During Doyle's tenure a series of territorial reannexations occurred that eventually tripled Providence's land area and brought the city to its present physical dimensions. In 1868 Washington Park, Elmwood, West Elmwood, and most of South Providence were reclaimed from Cranston; in 1873-74 the present North End, Wanskuck, Eagle Park, Elmhurst, Mount Pleasant, and Manton were acquired from North Providence; and in 1898 Hartford Park, most of Silver Lake, and a portion of Olneyville were annexed from Johnston. This dramatic physical growth placed great demands on the fire service, necessitating its numerical enlargement and the construction of stations in the newly acquired neighborhoods.

During the fifteen-year tenure of Chief Greene (1869-1884), a number of technical and administrative innovations took place. A more sophisticated and efficient telegraph alarm, the Gamewell system, was introduced in 1870, producing a phaseout of the old bell-ringing pro-

cedures by 1873. An ordinance of March 1883 created the position of superintendent of fire alarm telegraph, and Charles G. Cloudman began a distinguished twenty-five-year tenure in that office. Cloudman's first report (January 1, 1884) indicated the new system already consisted of 127 signal boxes connected by 120 miles of galvanized wire. Meanwhile, in 1881, telephone communication was established among the city's fire stations.

Chief Greene also introduced the department's distinctive badge (1871), established the first permanent hose company (No. 4, in 1872), and presided over the creation of a protective company commanded by Charles H. Swan. This unit, a revival of the old goods committee, was sometimes called "the fire insurance brigade." Its mission was to protect goods from damage by water and other threatened injuries; its apparatus, according to Greene's 1875 annual report, consisted of "oil-cloth coverings, chemical machines for the extinguishment of incipient fires, and other paraphernalia required by this service, with a suitable wagon for carrying the same, the expense of which is borne in part by the City, and part by the Providence Board of [Insurance] Underwriters."

In that year the department also placed in service a new Skinner extension-ladder truck in an attempt to cope with fires in the city's multistory buildings. The traditional seventy-five-foot extension ladders, hand-raised by nine men, were clumsy, heavy, unsafe, and increasingly inadequate. The Skinner, America's first practical apparatus-mounted aerial ladder, was patented several years earlier by George Skinner of New York. Its three wooden sections, mounted over the rear axle, telescoped when not in use. With a portable extention the Skinner reached heights of approximately one hundred feet.

The new device was assigned to Hook and Ladder No. 1, but it was regarded as flimsy and precarious, and its defects relegated it to the reserve list by 1884. On March 20 of that year the department acquired the much supe-

A native of Massachusetts, Dexter Gorton moved to Providence in 1845. Gorton was a carpenter by trade, and utilizing his skill and ambition to expand his business, he became a major building contractor. In 1863 Gorton was appointed second assistant engineer of the Providence Fire Department. Within two years he was first assistant, and on July 10, 1865, he was elected to succeed the ailing Charles Dunham as chief engineer.

During Gorton's four years as head of the department, dramatic improvements were implemented. In 1867 the Board of Engineers abandoned the last of the hand engines. A year earlier Gorton had convinced the City Council to appropriate thirty-four thousand dollars for the purchase of four new steam engines, four hose carts, and two hook-and-ladder trucks. Five new stations were constructed on South Main, Benevolent, North Main, Richmond, and Harrison streets to accommodate this equipment, and the department purchased its first horses to pull the apparatus.

Gorton stepped down from his post as chief engineer on June 10, 1869. His interest in the department, however, did not fade. As a member of the Providence Common Council between 1886 and 1888, the former chief served on the council's Committee on the Fire Department. When the Board of Fire Commissioners was organized in 1895, he was elected to its initial membership and remained on this powerful board for ten years. Gorton died in 1907, two years after his retirement, at the age of eighty-one. Engraved portrait by Ryder-Dearth, courtesy of Rhode Island Historical Society, Rhi x3 4970.

TELEGRAPH FIRE ALARM.

The TELEGRAPH FIRE ALARM will be used for signalizing all fires that occur within or near the city.

The bells to be used for alarming the Fire Department are upon the following buildings, viz.: ST. JOHN'S CHURCH, BENEVOLENT CONGREGATIONAL CHURCH, (Dr. Hall's, in the night,) ST. STEPHEN'S CHURCH, THIRD BAPTIST, (in the night,) ROGER WILLIAMS CHURCH, HIGH STREET CHURCH, (in the night,) NO. 9 ENGINE HOUSE, CENTRAL BAPTIST CHURCH, and G. T. SWARTS TOWER.

The signal boxes where alarm can be communicated may be found in all the Fire Engine Houses, (except No. 12,) in the Store of A. F. FULLER. No. 66 Wickenden Street; the POLICE OFFICE; the RESIDENCE of the CHIEF ENGINEER, No. 52 Meeting Street; the Store of BYRON SMITH, No. 203 North Main Street; G. T. SWARTS' HOUSE, 54 Pine Street; NEW MARKET BUILDING, corner of Broad and High Streets; and the STORE of WILLIAMS & SON, 273 High Street.

Upon the discovery of fire in any locality within or near the city, let the fact be made known to those having charge of the nearest signal box and the alarm will be immediately telegraphed through the line.

RULES

FOR OPERATING THE TELEGRAPH FIRE ALARM, ADOPTED BY THE BOARD OF ENGINEERS, JULY, 1862.

ARTICLE 1. The operators are the Police, Night Watch, Firemen, and those having charge of the boxes in the several Stores.

ART. 2. The first one of the operators who gets possession of a signal box, shall commence operations by making from five to ten rapid movements of the knob as preliminary signals. He shall then strike the number of the ward the fire is in, two minutes or more.

ART. 3. To avoid striking the wrong District, no bell ringer shall leave their station to ring their bell until they shall have received two or more distinct signals from the District through their box.

ART. 4. Upon receiving the correct alarm they shall proceed immediately to the bells assigned them, and after ringing smartly thirty seconds, shall toll distinctly the number of the ward where the fire is five times in succession, at slight intervals, and repeat the same five times at least.

ART. 5. It shall be the duty of all persons authorized to make signals for fire alarms, to note whether the armature is against the magnet or lying upon the pin; if in the latter position, then the telegraph is not in working order; the bell ringers shall then proceed to ring an alarm in the usual manner, and the firemen will use due diligence in finding the place of the fire if any there be.

ART. 6. The Foremen of the several companies will report immediately to the Chief Engineer when the alarm shall have been given from their station and the name of the person giving the alarm.

ART. 7. Those having charge of the signal box at Station No. 1, on Exchange Place, will every evening at 9 o'clock, strike the number of the hour. If the response is not made at the other stations throughout the city, it is required of the several operators to report the same at the office of the Chief, Room No. 11, Infantry Building, Dorrance Street.

ART. 8. Any employee of the Fire Department failing to comply with the above Rules, on being reported to the Board of Engineers, will be liable to a discharge from the Department.

CHARLES H. DUNHAM, *Chief Engineer.*
JAMES SWEET, *1st Assistant Engineer.*
DEXTER GORTON, *2d Assistant Engineer.*

PROVIDENCE, JULY, 1862.

Improvements in the method of reporting fires followed on the heels of the introduction of steam fire engines to the city. In 1860 the Providence City Council appropriated a thousand dollars to fund the installation of a telegraph fire alarm system developed by veteran fireman and councilman Charles E. Carpenter from a model devised in Boston by Dr. William F. Channing. A four-mile-long circuit of galvanized wire connected seventeen locations—the engine-houses, police headquarters, the room of the night watch, the chief engineer's office, and several stores whose proprietors had agreed, for a salary, to ring church bells upon receiving a signal of fire.

When a fire was discovered, an authorized person at any of the locations could set the alarm process in motion by tapping out the number of the endangered ward. The initial telegraph signals activated gongs in each of the other sixteen locations. Within two to three minutes bell ringers were at work tolling the ward number five times in succession to direct firefighters to the affected area of the city.

Rapid expansion of the city's population and the acquisition of a large area of land from Cranston made the system less effective. In 1870 Carpenter's fire alarm telegraph was replaced by the new Gamewell alarm system. Broadside, 1862, courtesy of Rhode Island Historical Society, Rhi x3 4963.

Providence's two great public works projects of the late nineteenth century were its waterworks and its sewer system. Both owed their origin to the initiative of Mayor Thomas Doyle and the expertise of engineer J. Herbert Shedd.

Preliminary studies for a water supply were conducted in 1868, and the Pawtuxet River was chosen as the most desirable source. Accordingly, construction of a reservoir on Sockanosset Hill in Cranston (180 feet above tidewater) and a pumping station at nearby Pettaconsett on the river's west bank was begun in May 1870. By December 1, 1871, thirty miles of iron pipe had been laid into Providence, and the first consumers drew Pawtuxet water from their taps. Though the Sockanosset Reservoir (for which Reservoir Avenue is named) had a 51 million-gallon capacity, an additional storage reservoir was needed almost immediately. Therefore the Board of Water Commissioners acquired a tract of land on the East Side between Hope, Olney, Brown, and Barnes streets and between 1873 and 1875 constructed the Hope Reservoir and pumping station (shown in this aerial view).

In 1889 a third "high service" reservoir and pumping station was completed at an elevation of 274 feet above sea level in the Fruit Hill section of North Providence just west of the city line. The availability of pressurized water resulted in the gradual phaseout of fire wells (cisterns) as a source of water. By the end of the century the Department of Public Works had installed almost eighteen hundred hydrants around the city. Photograph, courtesy of Rhode Island Historical Society, Rhi x3 4606.

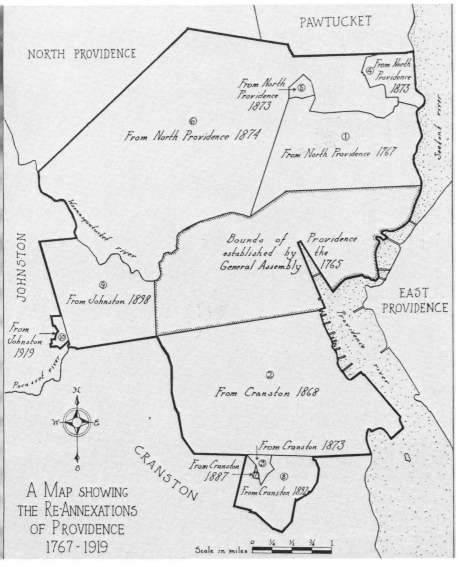

A MAP SHOWING
THE RE-ANNEXATIONS
OF PROVIDENCE
1767-1919

Scale in miles

The impetus for the extension of the Providence water supply system came not only from the growth of the city's population but from the expansion of its physical boundaries as well. During its Golden Age, Providence grew from less than six square miles to eighteen. This informative map drawn by John Hutchins Cady, the city's premier historian, details the annexation process. One reacquisition (which was more in the nature of an adjustment) took place in 1767 immediately following the creation of the town of North Providence. The final addition (from Johnston) occurred in 1919. The other eight annexations came between 1868, when a large chunk including South Providence, Elmwood, and Washington Park was taken from Cranston, and 1898, when Silver Lake, western Olneyville, and Hartford Park were reclaimed from Johnston. Map of Providence from John Hutchins Cady's Civic and Architectural Development of Providence.

In 1839 a young Harvard medical school graduate, Dr. William F. Channing of Boston, produced an invention that revolutionized the system of reporting fires. This was the telegraph firebox, a device designed to transmit to a central station the box number of an alarm activated by either cranking a handle or pulling a lever. This motion caused a notched gear to turn, thereby transmitting electrical impulses that opened and closed circuits at the central station. As soon as the box number was identified, dispatchers relayed the number by telegraph to bells in the fire stations.

Channing's fire alarm system was first adopted by Boston in 1852. Seven years later John N. Gamewell of Camden, South Carolina, purchased Channing's patents and began production of the Gamewell fire alarm system (right).

Providence adopted the Gamewell system in December 1870, when 50 street boxes were installed throughout the city. By 1900 more than 350 boxes were in service. Initially alarm boxes were locked with keys distributed to "responsible persons" living or working within the vicinity of a box, but the limitations of this arrangement eventually resulted in the adoption of keyless doors. The alarm system devised by Dr. Channing is still in use today, and Gamewell boxes remain an important weapon in preventing the spread of fire. Line Drawing from White's Fire Service of Providence, courtesy of Rhode Island Historical Society, Rhi x3 4945.

During the 1870s the fire department, with the encouragement of Providence Mayor Thomas Doyle, initiated an ambitious modernization program. The Wickenden Street station (below) was one of three constructed in 1875. Located on the corner of Wickenden and Traverse Street, the station was erected on the site of the old Gazelle volunteer fire company house. The large three-story brick building housed Hose Company No. 15 and Hook and Ladder Company No. 4. Both pieces of equipment pictured here in front of the station were built by Moulton and Remington of Providence.

The station was enlarged in 1895, with the east end serving as a precinct of the police department. In 1952 it fell victim to a twentieth-century modernization program and was closed. Ladder 4 was subsequently moved to the new station on the corner of North Main and Meeting streets, the former site of a Quaker meetinghouse. Photograph by J. W. Goodwin, courtesy of Rhode Island Historical Society, Rhi x3 4934.

Oliver E. Greene joined the Providence fire service in March 1854, following the establishment of the paid department. He worked as clerk for Fire King Hand Engine Company No. 5 for six years and then was assigned to Steam Engine Company No. 3 as its captain. Greene continued to move up the ranks, becoming second assistant engineer under Dexter Gorton in 1865. Four years later he was elected chief engineer.

During Greene's tenure as chief, massive improvements were made in the city's water system. On December 1, 1871, water from a pumping station on Sockanosset Hill in Cranston became available to the city's residents and its fire department. By the time of Greene's retirement in 1884, more than twelve hundred fire hydrants had been installed in the city, making the old system of pumping water from cisterns obsolete. During Greene's tenure the fire department also began an accelerated phaseout of the temporary call force. In 1870 only twenty-four men served as permanent department members, but by 1884 that number had more than tripled. Greene introduced the highly successful Gamewell system of fire alarm, and by the time he relinquished his office almost 150 signal boxes had been installed around the city.

One biographical sketch written shortly after Greene's retirement cited the thirty-year department veteran for being "a man of firm resolutions, a good organizer, and quite a disciplinarian." Greene retired to accept the office of sealer of weights and measures. He died on March 30, 1900. Photograph, courtesy of Rhode Island Historical Society, Rhi x3 4955.

rior Hayes ladder truck, manufactured by the La France Company of Elmira, New York, from designs patented by Daniel D. Hayes, a San Francisco Fire Department machinist. Hailed by fire buffs as the first successful aerial ladder, it could be raised to a height of eighty-five feet by a horizontal worm gear turned with a long handle that was cranked by as many as six men.

Other new devices from the Greene era included Babcock hand extinguishers (1873), sliding poles (1882), and two chemical engines made by the Babcock Fire Extinguishing Company of Chicago. These units, placed with Hose Companies No. 6 (Benevolent Street) and No. 7 (Richmond Street) in 1882, carried a full load of hose and two thirty-gallon chemical (soda and acid) tanks to extinguish small, incipient fires and chase sparks and embers.

Legislative changes also affected firefighting during this era. Most notable was the building code of 1878, the first comprehensive housing law since the pioneering statute of 1817. Among its thirty-eight sections were provisions requiring metallic fire escapes in all buildings employing "twenty-five or more operatives...in any of the stories above the second story"; at least one "metallic stand-pipe with hose coupling and hose or sprinkling pipes accessible from the main stairway" on each floor of any building over fifty feet in height; and "at least two egress stairways" in every building over two stories high "where combustible articles are manufactured and wherein over two hundred operatives are employed." This statute also made the chief engineer the inspector of buildings, but Greene was relieved of this duty by an April 1883 statute that created a separate office of building inspection.

Consideration was given to the safety of public buildings in 1882 by a joint special committee appointed by the City Council to examine the theaters and halls used for public gatherings with respect to strength of construction, fire hazard, means of egress, and ventila-

tion. The buildings examined were the Providence Opera House, the Music Hall, the Amateur Dramatic Hall, Infantry Hall, the Theatre Comique, Low's Opera House, the Central Congregational Church, the Union Congregational Church, and the Beneficent Congregational Church. In a report from experts retained by the committee, various recommendations were made for reducing the danger in case of fire, including the provision of additional exits and stairways, exit signs, and stair handrails; the fireproofing of arches and stairways; the protection of gas footlights by wire screening; and the installation of sprinklers and fire extinguishers. The owners of the buildings were notified by the committee and complied with its specific recommendations.

The joint special committee also urged the adoption of amendments to the building laws to provide for the safety of public buildings thereafter erected. It was not until 1895, however, that the desired amendments were passed by the General Assembly.

From time to time during the 1870s and 1880s, the rules and regulations of the department underwent revisions, and changes in the administrative structure occurred. Among the more notable rules was an 1878 provision declaring that "permanent members of the Department are not permitted to act as executive committeemen, or attend any political convention as delegates, or be allowed to hold tickets [ballots] at any public election." In addition, the rules then in effect banned "loud or boisterous talking" and "profane or obscene language" in or about the station houses and prohibited smoking in the sleeping quarters. Beds were to be made up before ten o'clock each morning.

The most notable administrative changes during the Greene regime, in addition to the creation of the post of superintendent of fire alarm, were the establishment by state enabling act of the position of fire marshal (1880) to investigate the cause of every blaze, with Elias M. Jenckes

On September 25, 1877, Providence Journal readers learned of the massive fire that had partially destroyed the U.S. Patent Office in Washington, D.C. The flames had incinerated thousands of models of early American inventions, as well as countless valuable documents and records. Two days later Providence suffered its own fire disaster, the worst since the Great Blaze of 1801.

At a few minutes after six in the evening of that day, someone noticed a flicker of fire in a building owned by Mrs. Nelson Aldrich and occupied by Waldron, Wrightman, and Company, wholesale grocers, and Charles W. Jenckes

and Company, paper box manufacturers. The building was located on the east side of Custom House Street at the corner of Pine. This four-story structure was in one of the most densely packed areas of the commercial district. By the time the first hose companies arrived, the front of the building was a wall of flame. Soon the Daniels Building across Custom House Street was also ablaze. Firemen positioned the Skinner ladder in front of the Wilcox Building on Weybosset Street and ran a hose to the roof of that structure to prevent the spread of fire to the west. But flames soon engulfed two more buildings, prompting Providence Mayor Thomas Doyle to

telegraph a plea for assistance to Boston, Fall River, Newport, and Pawtucket. Firemen retreated just before the front wall of the Vaughan Building collapsed.

Owners of threatened buildings dragged valuables to the sidewalks of the Crawford Street Bridge, and a detachment of the First Light Infantry was dispatched to protect these valuables from looters. Before the great fire of September 27, 1877 subsided, four buildings were completely destroyed, with a loss of almost $500,000. Photograph, September 29, 1877, courtesy of Rhode Island Historical Society, Rhi x3 4950.

PARK GARDEN,

PROVIDENCE, R. I.

◆

Thursday, October 10th, 1878.

◆

HAND ENGINES.

1st Prize, $200. 2d, $100. 3d, $75. 4th, $50.

Programme—Afternoon.

◆

		Ft.	In.
1.	Rhode Island, No. 6, Newport, R. I. - - J.		
2.	Cataract, No. 2, Franklin, Mass. - - H.		
3.	Washington, No. 1, Milford, Mass. - - B.		
4.	Eureka, No. 1, Arlington, Mass. - H. & D.		
5.	Narragansett, No. 2, Cedar Grove, R. I. - J.		
6.	Dog Island, No. 101, East Cambridge, Mass. S.		
7.	Mechanic, No. 2, Warren, R. I. - - - J.		
8.	Quinsigamog, No. 1, Hopkinton, Mass. - H.		
9.	Volunteer, No. 4, Peabody, Mass. - -B H.		
10.	Butcher Boy, No. 1, So. Braintree, Mass. - H.		

S—Stands for Smith; **H**—Hunneman; **B**—Button; **J**—Jeffers; **H & D**—Howard & Davis; **L**—Leslie, Makers.

Official Order of Play,

IRVIN BESSE, Sec'y.

Thursday Evening.

◆

8.00 o'clock. Concert by AMERICAN BAND.

8.30 o'clock. Torchlight Procession in the Garden.

9.00 o'clock. FIREMEN'S RACES. One Mile race—Two Prizes. Three-legged Race—One Prize. Wheelbarrow Race—One Prize.

9.30 o'clock. GRAND FIREMEN'S BALL.

10.00 o'clock. BRILLIANT ILLUMINATIONS,

AND MOUNT VESUVIUS.

7-20-19

Firemen's musters served to perpetuate the traditions and nostalgia of the old volunteer fire service. One such muster took place at the Park Garden on Broad Street on the evening of October 10, 1878. On that day several hundred firemen from Rhode Island and nearby Massachusetts marched from Exchange Place to the muster grounds. Using water from an artificial pond at the Park Gardens, participants strained on the old hand pumpers in an attempt to project a horizontal stream of water over a 6-foot-wide plankway covered with thick brown paper that extended 230 feet away from the competitors.

Two thousand spectators watched the Quinsigamog Fire Company from Hopkinton, Massachusetts, take first honors by pumping a stream of water 178 feet long. That evening the "Three Ones" of Providence sponsored a firemen's ball at Howard Hall in Providence. Both the annual ball and firemen's muster have remained time-honored traditions among firefighters throughout the United States. Program, October 10, 1878, courtesy of Rhode Island Historical Society, Rhi x3 4939.

These two photographs of the Olneyville Fire Department were taken around 1880. Dubbed "Rough and Ready," Eagle No. 2 occupied a station house on the corner of Plainfield and Rye streets (right) that was located in the town of Johnston until the annexation of 1898. In that year the structure became the home of new Hose Company No. 19 (left). The old Eagle hook-and-ladder truck (lower left) was rebuilt by John G. McIntosh of Providence and returned to service on July 19, 1898.

Hose 19 was housed on this site until 1950, when the station (right) was abandoned. A year later the building was sold via city auction to the Regal Card and Printing Company. Photographs of Olneyville Fire Department, circa 1880, courtesy of Rhode Island Historical Society, Rhi x3 4448 and 4931. Photograph of Hose 19 station, 1931, courtesy of Rhode Island Historical Society, Rhi (B877) 1063.

the first incumbent; the creation of the post of deputy chief engineer (1882), with George A. Steere the initial appointee; and the founding (1881), by eighty-five-year-old Zachariah Allen, Elisha Dyer, Jr., Stillman White, Charles E. Carpenter, and others, of the heritage-oriented Providence Veteran Firemen's Association, an organization to collect "for permanent preservation all records, papers, documents, legends, memorials, and relics relating or pertaining to the volunteer or paid department of the city."

When Chief Greene retired in July 1884, after thirty years of service, the fire department consisted of the Board of Engineers, the superintendent of fire alarms, and a manual force of 176 men. These 176 were divided into two classes—permanent members (78) and call men (98). The call men were those who engaged in fire service when the bell alarm was sounded rather than on a continuous and permanent basis; in effect, they were temporary, ad hoc, part-time employees, and Greene had recommended that they be phased out. In the year of his accession, 1869, there had been 24 permanent and 93 call firefighters. Obviously the department's growth had occurred principally in its permanent force.

The roster for 1884 listed five steamer companies: Niagara No. 5 (North Main Street, Pawtucket-made Cole Brothers engine); Atlantic No. 8 (Harrison Street, Silsby engine); Washington No. 10 (Burnside Street, Silsby engine); Stillman White No. 12 (Smith and Orms streets, Cole Brothers engine); and Putnam No. 14 (Putnam Street, Cole Brothers engine). In addition to these glamour units, there were ten hose companies, one private hose unit, two chemical engine companies, the protective company, and four hook-and-ladder units, one with a new Hayes

aerial, two with trucks built by Ryan Brothers of Boston, and one with a truck built by Moulton and Remington of Providence. Horses in active service to pull this apparatus numbered forty-three.

In 1884 there were 1,218 hydrants tied in to the Providence water supply system, but forty-nine wells and cisterns remained in use. The alarm system consisted of 130 signal boxes, 130 miles of wire, and eighteen telephones.

The deparmental budget was $111,074, apportioned as follows: salaries, $84,704; new apparatus and hose, $6,593; telegraph, $3,935; and general expenses, $15,842. The latter included such items as veterinary fees (to Dr. A. C. Buchanan, who served the department for thirty years), hoof ointment, mare and tail combs, hay and grain, shoeing, lanterns, and bedding.

Despite the improvements in fire service and fire safety, several fatalities and disastrous fires occurred in Providence during the last third of the nineteenth century. Two firefighters were victims of the new technology. In September 1870 a Jeffers steam fire engine No. 6 exploded en route to a fire on East Street, fatally injuring a bystander and contributing to the death of assistant engineman John H. McLane; in October 1883 the department's Skinner ladder snapped during a practice drill on Exchange Place, plummeting thirty-year-old Alexander J. McDonald ninety feet to his death. To memorialize these and other deceased firefighters, a thirteen-foot Westerly granite and bronze monument was erected on the firemen's burial lot in the North Burial Ground and dedicated on August 16, 1885, with addresses by Mayor Doyle and the Reverend Charles L. Goodell.

The most notable late nineteenth-century fires

This engraving appeared in Frank Leslie's Illustrated Magazine on November 19, 1881. The city of Providence played host to a French delegation led by Marshal Petain, which was in America to commemorate the centennial of the defeat of British forces at Yorktown. The maneuvers of the Providence Fire Department, staged for the European visitors, drew large crowds to the Crawford Street Bridge area.

The sound of alarm box 312 brought seventeen pieces of equipment, with the first piece arriving only one minute after the alarm call. The department's Skinner truck is pictured here being raised against the Day Building.

This demonstration, the Providence Journal commented, "was only another proof of the admirable discipline and system which dominate in this department." Engraving, courtesy of Rhode Island Historical Society, Rhi x3 4928.

The "Three Ones" fire station was opened in 1875 to replace the old Hydraulion 1 station that had stood on the northwest corner of Exchange Place for the previous half century. The new two-story central station was constructed on pilings directly over the confluence of the Woonasquatucket and Moshassuck rivers. Headquartered there were Hose No. 1, Hook and Ladder No. 1, and the Protective Company. Office space on the second floor was reserved for the chief engineer, the superintendent of lights, the overseer of bridges, and the public buildings department. The Three Ones served the city until 1903, when the new Central Fire Station was opened just north of this site. Photograph, circa 1890, courtesy of Rhode Island Historical Society, Rhi x3 4536.

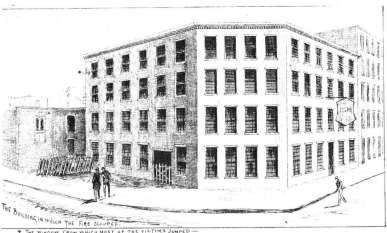

THE BUILDING IN WHICH THE FIRE OCCURED.
* THE WINDOW FROM WHICH MOST OF THE VICTIMS JUMPED —

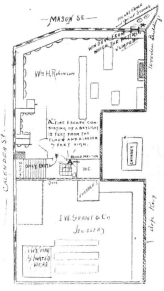

ROOM WHERE THE FIRE STARTED. "LE JOLLY" DYE HOUSE.

ROOM IN WHICH THE GIRLS WERE AT WORK — SHOWING THE ONLY DOOR OF EXIT

VIEW FROM THE DOOR OF THE ROOM IN WHICH THE FIRE SMOKE ETC.

One of the most tragic fires in the city's history occurred on the morning of November 21, 1882. For the workers at a four-story brick factory on the corner of Calendar and Mason streets, that day began much like any other. On the fourth floor were two jewelry shops. The shop at the north end of the building employed about forty persons, mostly young women. One floor below them was the Le Jolly Dye House. The owner, Charles T. Melvin, was in the process of cleaning a garment with naphtha. Nearby a plumber was using a vapor stove to repair a large drying cylinder. Suddenly a spark from the stove ignited the naphtha. Unable to check the flames, Melvin left the room to sound the alarm.

Upstairs, jewelry workers at the south end of the building heard the shouts of fire and retreated down the one stairway to safety. For the workers at the north end, it was too late. Smoke and flames drove them back when they attempted to descend the stairs. Because the building lacked fire escapes, their only alternative was to jump. The group moved to the far end of the factory in close proximity to an abutting two-story building. They decided to open the window and leap one by one to the roof of the adjacent structure twenty feet below. But panic seized the workers, and more than a dozen, pushed from the window, plunged fifty feet to the ground. Two

women fell headlong into a pile of grindstones directly below the window; another was impaled on a picket fence. By the time the flames were extinguished, four workers were dead and sixteen seriously injured. Ironically, the flames were extinguished quickly, with only five thousand dollars in damage to the building.

A coroner's inquest cited the owners for failing to provide adequate means of escape. Ensuing public outrage prompted the passage of a law by the Rhode Island General Assembly requiring the installation of fire escapes in most commercial and manufacturing buildings.
Series of line drawings, courtesy of Rhode Island Historical Society, Rhi x3 4920.

A *year after the Calendar Street tragedy, the fire department suffered a fatality in its own ranks. On the evening of October 4, 1883, Captain John C. Tuttle of Hook and Ladder Company No. 1 had positioned the Skinner truck (also known as Blue Pointer) in front of the Three Ones station for a training drill. Thirty-year-old Alexander J. McDonald, a second ladderman and two-year veteran of the force, quickly scaled the ninety-foot ladder, closely followed by another fireman. When McDonald reached the top, the upper section slowly began to sway in the direction of the railroad depot. The ladder snapped, sending McDonald crashing into the cornice of the station, killing him instantly. The second fireman grasped for a guide rope used to support the ladder and swung to the ground, receiving only slight injuries.*

McDonald's funeral brought firemen from as far away as New York. The two-division procession from Exchange Place to the North Burial Ground was witnessed by large crowds that turned out not only to express sympathy for McDonald's widow and child but also to recognize the distinguished service provided by the Providence Fire Department. McDonald's untimely death prompted a campaign to build a permanent monument to those firemen who risked their lives to ensure the safety of Providence residents. Engraving from White's The Providence Firemen, *courtesy of Rhode Island Historical Society, Rhi x3 5101.*

Shortly after McDonald's death the Providence Association of Firemen launched a fund-raising campaign for a monument to be placed on a twelve-hundred-square-foot plot of land that the association owned in the North Burial Ground. By April 1884 more than twenty-two thousand dollars had been collected. The association also sponsored a public design competition, which was won by Frank Tingley of Providence, at that time the state's foremost monument designer.

The monument, fashioned from Westerly granite and bronze, stands thirteen feet six inches high. The figure atop the base stands seven feet tall, and its features are said to resemble those of the slain fireman. A copper box buried beneath the cornerstone contains several articles from local newspapers, a copy of the annual report of the chief engineer for 1884, a copy of the constitution of the Providence Association of Firemen for Mutual Assistance, and all correspondence relating to the dedication, which was held on August 16, 1885. Photograph by Denise Bastien.

included the September 27, 1877, conflagration that destroyed the factory of C. W. Jenckes on Harkness Court and Custom House Street, causing a loss estimated at $500,000, and the Calender Street fire of November 21, 1882, killing four factory workers and injuring scores of others who jumped to safety.

Two of the city's largest department stores went up in flames during the Christmas shopping season in 1890. On December 5 the Shepard Company sustained a $200,000 fire loss, and one week later the J. B. Barnaby Company at Westminster and Dorrance was totally destroyed in a devastating $400,000 blaze. In February 1896 the original and vacated Union Train Station on Exchange Place, an architecturally significant structure and the country's largest depot when built in 1849, was gutted, with a loss of $50,000; one month later the original Masonic temple on Dorrance Street was destroyed in a $260,000 fire.

The most significant fire losses of the era came during one disastrous week in mid-February 1888. Within a span of eighty-eight hours (February 15-19), the Aldrich House Hotel on Washington Street and the entire city block surrounding it (presently occupied by the Biltmore parking lot and garage) were leveled, with a loss of $400,000; the popular Theatre Comique on Weybosset Street burned ($100,000); and the Daniels and Cornell Building on Custom House Street was destroyed ($175,000). With this devastating start, fire losses in 1888 totaled $750,000, the highest of any year of the century.

On July 2, 1884, George A. Steere took command of the department upon the retirement of Oliver Greene. Having begun his career in 1861 as a member of the Gaspee Hand Engine Company, Steere advanced to the top post from the job of deputy chief engineer, a position created in 1882.

Steere headed the fire service until April 1909, just three months short of twenty-five years. The first part of his tenure was notable more for political and administrative activity than technical innovation. Steere had interrupted his fire service to represent his West End neighborhood on the Common Council between 1877 and 1879, and that connection had undoubtedly influenced his selection by that body as deputy chief and then department head.

Shortly after Steere's accession the council assumed a more active role in directing the fire service. In March 1885 it adopted an ordinance abolishing the Board of Engineers and transferring the board's power to the chief engineer and the council's Joint Standing Committee on the Fire Department, where firemen-turned-politicians Ira Winsor and Dexter Gorton were the dominant figures. Then, in 1890, the council provided the chief with a three-year term.

Steere also used his political instincts to great advantage in 1886, when Providence—as part of its mammoth 250th birthday celebration—hosted the International Association of Fire Chiefs. At that group's August convention Steere was chosen president of this prestigious professional organization.

In the 1890s the fire force became embroiled in political controversy reminiscent of the last days of the volunteer era. Again, increased public spending was part of the problem. In the decade from 1885 through 1894 the department's annual budget more than doubled, from $150,000 to $304,000. Charges of political favoritism in the selection of men for the permanent force—which increased from 96 to 215 during the same period—also prompted demands for the creation of an administrative commission to run the department.

In May 1892 the General Assembly authorized such an agency, though it disregarded the wishes of reformers by providing for council rather than mayoral appointment of these commissioners. Both the council and its critics were dissatisfied—the former wanted no change at all; the latter desired a further curtailment of council influence—and so a long impasse occurred, prompting the city solicitor to observe in December 1893 that "in the absence of the appointment of the commission required by law, there was no lawful control over the department of any kind or any power in any person whatsoever to enforce discipline or give any lawful orders or directions."

This confused state of affairs continued until February 1895, when the council relented and complied with the 1892 state statute. On February 27 a salaried three-man commission consisting of Stillman White, William H. Luther, and former chief Dexter Gorton took office, vested with the complete control and management of the

In addition to his duties as foreman of Hook and Ladder Company No. 1, Captain Charles E. White spent a great deal of time chronicling the history of the Providence Fire Department. In 1886 he published a 313-page illustrated history of the department entitled The Providence Fireman and followed this a year later with Fire Service in Providence. The second book was published as a fund-raiser for the Providence Permanent Firemen's Relief Association.

White left the department shortly after the second book was published and died in Providence in 1899 at the age of forty-six. Engraved portrait by Ryder-Dearth, courtesy of Rhode Island Historical Society, Rhi x3 5154.

This photograph of the Three Ones was taken in June 1886 during the city's observance of its 250th anniversary. Photograph, 1886, courtesy of Rhode Island Historical Society, Rhi x3 4995.

The Providence Veteran Firemen's Association was established in 1881 "for the purpose of collecting for permanent preservation all records, papers, documents, legends, memorials and relics relating or pertaining to the volunteer or paid fire department of the city," and to encourage "professional fraternity for mutual benefit and pleasure." Eighty-four-year-old Zachariah Allen was elected the association's first president. The Providence veterans pictured here were fre-

quent competitors in firemen's musters and sponsored a number of social events. The group remained active for more than six decades, finally disbanding in the mid-1940s. Group portrait, circa 1890, by Arthur S. Adams, courtesy of Providence Fire Department. Photograph of the association's helmet and belt by Denise Bastien, courtesy of Museum of Rhode Island History, Rhode Island Historical Society.

fire department, including the power to appoint and discharge all of its members, to purchase new apparatus, to sell the unserviceable equipment, and to fix the salaries of the firemen and officers subject to the approval of the City Council.

In the aftermath of this political controversy, the commissioners, chaired by White, attempted to depoliticize the department. In November 1895 they issued an order forbidding any partisan political activity by firemen. Then they gradually phased out the call men, from forty-three in early 1895 to none by January 31, 1899, when the remaining fourteen were granted "an honorable discharge." Thenceforth the department consisted wholly of permanent men. According to General Orders No. 2, these employees were prohibited from performing any other kind of work or using any of the fire department's premises for the purpose of transacting business unrelated to the fire service. Eventually, in 1901, the board established competitive civil service examinations as a basis for promotion.

In 1887, under Steere's leadership, the Providence Permanent Firemen's Relief Association was formed. This group was similar. to the Providence Association of Firemen for Mutual Assistance, which had existed since 1829 to

provide death and disability benefits for members of the fire department and their families. The new organization was limited to full-time firefighters, and by 1899 it supplanted its older counterpart. An initiation fee of $3.00, plus $1.00 per month for the first year and $.50 per month thereafter, entitled a member's widow or heir to a $1,000 death benefit, and a small annual sum was provided to those placed on the retired list for "sickness, disability or age."

Another evidence of Steere's concern for firefighters' benefits is contained in his 1887 annual report, wherein he successfully pleaded with the council to allow permanent members of the force an annual vacation of one week. His statement of existing work conditions speaks for itself: "These men are now compelled to remain on duty every day in the year, Sundays included. True, they are allowed two days' leave of absence from the stations each month, but it must be understood that even then they are required to respond to alarms; nor are they allowed these two days in succession.... This is the only occupation where men are compelled to devote twenty-four hours each day to their duties." Who said the "old days" were good?

Although the volunteer fire companies were formally disbanded in 1854, the spirit and enthusiasm of their members did not fade. Several units, such as the Water Witch (or "Sixes") and the Fire King (or "Threes"), formed associations to perpetuate the traditions of the volunteer companies. Their activities were usually social in nature, although occasionally the groups would engage in target practice with their antiquated fire equipment. Here members of the Providence Water Witch Association pose in front of their engine during the group's annual outing at Rhodes-on-the-Pawtuxet. Photograph, circa 1885, courtesy of Rhode Island Historical Society, Rhi x3 4930.

Administrative changes were much more significant than technical innovations during the first decade of the Steere regime. For example, the much-acclaimed, sixty-five-foot-high "Champion" water tower, purchased in 1893, was adjudged "a useless piece of apparatus" by the commissioners in their 1899 report, which recommended the tower's disposal.

The major and most enduring improvements of the 1890s were the introduction of combination wagons (hose and chemical), the first being placed in service with Hose Company No. 18 (Broad and Rugby streets) on November 30, 1895; the passage of a statute in 1898 authorizing fire department inspections of all buildings other than private dwellings "where waste material of a combustible nature has been allowed to accumulate"; and, especially, the establishment of a high-pressure water system for the Downtown area. This "separate system of water service," as the commissioners called it, was completed on October 12, 1897. Initially there were eighty-nine hydrants, with pressures ranging from 90 to 120 pounds, connected to the system, which emanated from the Fruit Hill (or High Service) Reservoir located in North Providence 274 feet above sea level. Describing the new supply, the department proudly boasted that "the service, unlike any other

in the country . . . , is ready for immediate use the moment the hydrant head is connected, and does not depend upon the use of a fire boat or forcing pump. . . . It is one of the most valuable additions to our fire extinguishing facilities that has been made since the introduction of Pawtuxet water in 1871." In the years to follow, the high-pressure hydrants, together with the high-rise buildings, gave hose-and-ladder companies increasing importance within the firefighting service.

The fire commissioners took their final department inventory of the nineteenth century on December 31, 1899. This inspection yielded the following statistics on the fire service of Providence: 28 companies (8 steam-engine, 12 hose, 7 hook-and-ladder, and 1 protective); a force of 248 permanent men, equipped with the latest apparatus and housed in clean, spacious quarters; 88 horses; 33,222 feet of hose; 1,783 regular and 92 high-pressure hydrants; 351 fire alarm boxes (plus 18 alarm bells, 46 gongs, and 39 tappers); and an annual expenditure of $345,807 (up from $157,832 a decade earlier). The department had developed dramatically from the frantic bucket brigade that had battled the Great Fire of 1801, but even these improvements would be dwarfed by the technology of the century which lay ahead.

Daniel D. Hayes, a San Francisco Fire Department machinist, patented the first successful aerial ladder. In 1884 Providence purchased a Hayes truck from the La France Fire Engine Company of New York. The ladder was raised by a horizontal worm gear that was turned by a long crank operated by as many as six men. Within two months after its arrival in Providence, it was placed in service to fight a blaze at the Oliver Johnson and Company building. A fire was discovered in the three-story building on the corner of Exchange Street and Exchange Place at seven-thirty in the evening of May 26, 1884. Quick work by the Providence Fire Department—with the aid of their new Hayes truck—saved the structure, the only casualty being a cat owned by one of the occupants. On the day following this successful debut, the Providence Journal reported that the fire had been caused by "the traditional pipe in the Irishman's pocket." Engraving, courtesy of Rhode Island Historical Society, Rhi x3 4944.

Federal Hill marked the city's 250th anniversary by receiving a new firehouse. Located on the corner of America Street and Atwells Avenue, the 2½-story structure became the home of Hose Company No. 9 (which was transferred from its Pallas Street station) and new Hook and Ladder Company No. 6, equipped with a modern La France ladder truck. The station served that densely populated area of the city until 1951, when a new facility was opened at 630 Atwells Avenue. Photograph, 1886, courtesy of Rhode Island Historical Society, Rhi x3 4949.

55

George A. Steere holds the distinction of having held the position of chief longer than anyone in the history of the department. He began his career in 1861 as a member of the Gaspee Hand Engine Company. Five years later he was transferred to Steam Engine Company No. 3, and in 1875 he was appointed captain of new Hose Company No. 9 on Pallas Street. Dabbling in politics briefly, the Providence fire captain represented the Broadway area in the Common Council between 1877 and 1879. In 1882 he was appointed to the newly created position of deputy chief engineer, and when Oliver Greene resigned as chief on July 2, 1884, Steere was the heir apparent. He inherited a force of seventy-eight permanent firemen and ninety-eight ''call men,'' who operated five steamers,

four hook-and-ladder trucks, and an assortment of other equipment. When Steere retired in 1909, three months short of twenty-five years as chief, the equipment available to fight the city's fires had tripled. Steere oversaw the elimination of the ''call'' (part-time) firemen, thereby effectively increasing his force fourfold.

During his long tenure fourteen new stations were built, including the massive Central Fire Station, completed in 1903. In 1886 Steere became the first of two Providence firefighters elected to the presidency of the International Association of Fire Chiefs. He died on January 7, 1911, only two years after retiring from a career that spanned nearly half a century. Photograph, courtesy of Rhode Island Historical Society, Rhi x3 4969.

Local manufacturers of fire equipment played a prominent role in outfitting the fire department in the nineteenth century. Burr and Shaw of 46 Westminster Street began selling engine hose, fire buckets, and saddle equipment as early as 1840. Ephraim Moulton and Benjamin Remington, who began their business in 1849 as carriage painters, became major suppliers of trucks and wagons to the Providence Fire Department following its organization as a paid service in 1854 (right). By 1886 virtually all the department's hose carts and two of its six ladder trucks were Moulton and Remington's handiwork. The company continued producing fire equipment until 1899, when it was absorbed by the Armstrong Carriage Company.

The city's experience with the production of steam fire engines was less successful. Between 1875 and 1877 the Allen Supply Company

produced three steamers and then discontinued production. The Providence-based International Power Company—created by the merger of the Rhode Island Locomotive Works, the Corliss Engine Company, and the Greene-Wheelock Engine Company—acquired the rights of the famous Amoskeag Company of Manchester,

New Hampshire, to build steam fire engines, and it constructed at least fifteen Class C Amoskeags in Providence between 1907 and 1913. The city purchased one of these machines in October 1907 for Steam Engine Company No. 5. Advertisement from Providence City Directory, 1887, Rhi x3 4991.

The Combination Ladder Company of Providence began operations at 366 Fountain Street in 1887. Proprietor C. N. Richardson's Providence Fire Extension Ladders gained wide acceptance among fire departments throughout the United States and Europe. The company provided a variety of fire department supplies and were manufacturers of Seagrave trussed trucks. Pictured (left) is a combination chemical and hose wagon produced by the Combination Ladder Company for the Waterbury, Connecticut, fire department in 1906. Richardson's company produced fire equipment until 1960, when the company was absorbed by the Harris Lumber Company. Photograph, courtesy of Rhode Island Historical Society, Rhi x3 4449.

Grinnell, an internationally known manufacturer of fire protection systems, originated in 1850 as the Providence Steam and Gas Pipe Company. Initially it specialized in municipal gas work and industrial heating. During the Gilded Age the company diversified its operations under the leadership of Frederick Grinnell, a New Bedford native who had worked as superintendent of the famed Corliss Steam Engine Works prior to assuming the presidency of Providence Steam and Gas Pipe in 1869. Grinnell, who held a civil engineering degree from Rensselaer Polytechnic Institute, soon became acquainted with an automatic sprinkler device exhibited by Henry Parmelee of New Haven in 1874. Grinnell improved this sprinkler and in 1878 secured the rights to manufacture and install it on a royalty basis. In 1881 Grinnell himself patented a heat-sensitive automatic sprinkling device so advanced that one insurance organization recommended it by name to all its business clients. Working in concert with

the insurance industry, Grinnell was a major force in the creation of the National Fire Protection Association (1896).

In the years prior to his death in 1905, Grinnell continually improved his fire protection devices, acquiring dozens of patents in the process. By the end of the nineteenth century, his systems were in use throughout the country and in Canada, Europe, India, and Australia as well.

Influenced by the industrial consolidation movement of the era, Grinnell's business merged with an Ohio and a New York firm in 1892 to form the General Fire Extinguisher Company. Its new headquarters and plant on West Exchange Street are shown in this 1899 photo.

The great demand for the corporation's products—which also included power piping, valves, fittings, and hot-water heating systems—prompted the opening of a new foundry in 1909 on Elmwood Avenue in Cranston. Under the able and long-term presidencies of Frank H. Maynard (1906-1925) and Russell Grinnell

(1925-1948), the company continued to expand and prosper, changing its name in 1944 to the Grinnell Corporation. The consolidation movement of the modern era produced a merger of Grinnell with ITT in 1971. Because it also owned the Hartford Fire Insurance Company, ITT was soon ordered to divest itself of Grinnell's Fire Protection division. That portion of the Grinnell enterprise was bought by Tyco Laboratories in 1975 and presently operates from headquarters on Dorrance Street, while maintaining research and development operations at its Elmwood Avenue plant. In 1981, just one century after Frederick Grinnell's path-breaking patent, this new corporate entity—Grinnell Fire Protection Systems Company, Inc.—recorded another milestone in its long tradition of innovation and ingenuity by producing the first residential sprinkler to pass all Underwriters Laboratory tests successfully. Photograph, 1899, courtesy of Rhode Island Historical Society, Rhi x3 5003.

The most disastrous week in the fire annals of Providence began on the night of February 15, 1888. Just before midnight a locomotive engineer at Union Station noticed a glow emanating from the windows of a four-story wooden building on the corner of Eddy and Fountain streets. By the time the fire engine companies from the Three Ones arrived, the building was completely engulfed. The flames soon leaped across Worcester Street, setting the Billings stables on Union Street ablaze. Winds gusting to near fifty miles an hour and temperatures approaching zero made fighting the flames especially treacherous. At 12:20 A.M. the venerable Aldrich House hotel (now the site of the Biltmore Garage) fell victim to the advancing conflagration. By one o'clock the entire two-block area between Fountain, Eddy, Union, and Washington streets was involved. Sparks from the blaze were swept as far south as Rhode Island Hospital and ignited secondary fires on Richmond, Pine, and Broad streets. Firemen threw up a wall of water along Washington Street to prevent any further spread of the flames. Slowly the blaze was brought under control. By daybreak the surrounding streets were packed solid with ice. In some areas along Washington Street, hose was submerged under a foot of ice. The Hayes extension ladder of Company No. 6 was frozen solid against the wall of the Aldrich House. Though nineteen buildings were destroyed, miraculously no lives were lost.

Two nights later Providence firefighters were called to action once again. Shortly after midnight smoke was seen billowing from the roof of the Theatre Comique, located on the corner of Weybosset and Orange streets diagonally across from the Arcade. (This eighteen-year-old building had originally housed a billiard hall that quickly became "the resort of all the leading sporting men of the city," but a change in ownership resulted in the structure's conversion to a variety and burlesque theater.) Flames quickly spread to the front stage area. Chief George Steere called a second alarm, fearing that the fire in the flimsy wood-frame structure would trigger a repeat of the previous Wednesday's disaster. Fortunately, quick action by the department and a light wind limited the blaze to two buildings. The theater was a total loss. The

current attraction, ``Lilly Clay's Burlesquers,'' lost its costumes to the flames, but in typical show-biz fashion it appeared the following evening at Low's Opera House.

Less than forty hours later the infamous Alarm 121 sounded again. The Daniels Building on Custom House Street, constructed on the burned-out foundations left by the great blaze of September 1877, was engulfed in flames. Firemen moved into the interior of the building but were soon forced back by the intense heat. Eventually a second and a third alarm were sounded, committing the entire Providence force of 224 men to the blaze. The upper floor of the five-story building, housing the offices of printers J.A. and R. A. Reid, was completely destroyed. Mild weather aided Providence firemen in saving the structure, however, though it sustained damage amounting to approximately $100,000.

That year, 1888, the department reported fire losses totaling $750,000—the largest toll in the city's history prior to the twentieth century. Photograph of Aldrich House fire (left), February 16, 1888, courtesy of Rhode Island Historical Society, Rhi x3 4532 and x3 4935. Photograph of Theatre Comique (top right), February 19, 1888, courtesy of Providence Public Library. Photograph of Daniels Building (above), February 20, 1888, courtesy of Rhode Island Historical Society, Rhi x3 4994.

One of the most stubborn fires in the city's history began about 2:45 A.M. on Sunday, February 3, 1889. When firemen arrived at the Providence Coal Company on Dyer Street, they found a 250-foot-long wood storage shed ablaze. The inferno quickly drew thousands of spectators, who watched firemen fight to contain the flames. Within two hours the fire was under control, but the heat from the blaze had ignited several thousand tons of coal housed in the shed. Firemen remained at the site for the next three months, dousing the smoldering coal. Finally, on May 7, Chief Steere recalled the last engine company from the scene. Photograph, February 3, 1889, courtesy of Rhode Island Historical Society, Rhi x3 2555.

A familiar scene on the streets of American cities during the late nineteenth century was the horse-drawn steamer racing to a blaze, with the chief in his buggy alongside. The setting for this famous and oft-reproduced painting by Ernest Opper is Elizabeth, New Jersey, but the picture now hangs in Providence at the gallery of the Rhode Island School of Design. Photograph of painting, courtesy of Rhode Island School of Design.

INTERIOR OF PROVIDENCE ENGINE HOUSE.

READING ROOM.

SLEEPING ROOM.

STEAMER TEN RESPONDING TO AN ALARM.

The fire station of the Victorian age contained amenities designed to make life easier for those on the permanent force. Some houses, like the one depicted here, contained reading and billiard rooms. In this era, when department members were allowed only brief daily breaks for meals with their families, and no more than two twenty-four-hour periods away from their station each month, the station house was the firefighter's home. Engravings, courtesy of Rhode Island Historical Society, Rhi x3 4940-43.

The Christmas season of 1890 proved disastrous for Downtown retailers. On December 5 fire broke out on the first floor of the Shepard Company building on Westminster Street. Quick work by the fire department, however, confined the fire to that one level. More than $150,000 in dry goods was destroyed, but damage to the building was negligible.

Less than two weeks later the Westminster retail strip was hit by a second major conflagration. The J. B. Barnaby Company, on the southwest corner of Westminster and Dorrance streets, had become one of the most popular shopping emporiums in New England. On Saturday afternoon, December 13, the store was jammed with several hundred patrons purchasing gifts for the Christmas season. Just before three o'clock smoke was discovered pouring from the basement by one of the store's employees. Barnaby's general manager, H. B. Winship, attempted to coordinate an evacuation as thick smoke began to surge up through the stairway. Fortunately, recent state fire-escape laws prevented a repeat of the Calendar Street tragedy of 1882, and with the aid of fire department personnel the smoke-filled store was evacuated without a single fatality.

By 3:20 P.M. flames burst through the roof of the four-story building. Chief Steere called out the entire department and summoned help from five nearby communities. Just after four o'clock the Dorrance Street side of the building began to sway; moments later it collapsed, burying Ladder Truck 7 and injuring several firefighters. More than twenty-five firemen were hurt fighting this spectacular blaze. By six o'clock that evening the building was a complete loss.

With characteristic retail zeal the Barnaby Company recovered quickly, and less than ten months later a new four-story Barnaby store opened on the site of its unfortunate predecessor. Photograph, December 13, 1890, courtesy of Rhode Island Historical Society, Rhi x3 4533.

In 1892 a three-man Board of Fire Commissioners was authorized by the General Assembly in order to lessen City Council influence over fire department policy and patronage. Political resistance delayed the creation of this governing body until February 1895, however, when the council finally relented and designated three commissioners, each receiving an annual stipend of eight hundred dollars, to set all department policy. Shown here are the first appointees (left to right): John W. Morrow, clerk; Commissioner Dexter Gorton, building contractor and former chief engineer; Commission Chairman Stillman White, a prominent businessman; and Commissioner William R. Luther. The department was governed by this commission until 1931.

In 1895, the year of this major reorganization, privates in the department were divided into four grades and received between two and three dollars per day. Their duties included grooming the department's eighty-seven horses, cleaning harnesses, and maintaining such varied apparatus as steam engines, ladder trucks, ladders, hose, and chemical wagons. Snow added to their tasks, requiring the cleaning of the city's 1,648 hydrants as well. Photograph, 1901, courtesy of Rhode Island Historical Society, Rhi x3 5040.

During the winter of 1896 Providence firemen were put to the test by two spectacular blazes in the Downtown area. Shortly after midnight on February 21 fire broke out in the old Union Depot on Exchange Place. The mammoth 625-foot-long station had been a landmark in the Downtown since 1848, but plans had been completed for the construction of a new union railroad station on a knoll just to the north of the depot. City officials were in the process of considering new uses for the old station when fire decided its fate. Flames burst from the northeastern end of the building and quickly spread to the center of the structure. By the time the fourth alarm was sounded, the entire building from end to end was engulfed. Every piece of fire equipment in the city was committed to fight the inferno. By dawn Thomas A. Tefft's architectural masterpiece was a smoldering shell. Eventually work crews cleared the site and installed a small public park.

Less than a month after the depot fire, during the early morning of March 19, Providence firemen were again summoned to fight a major blaze. They arrived at the corner of Pine and Dorrance streets to find smoke billowing from the roof of the five-story Masonic Temple, an 1885 structure that was home to several Masonic lodges. For more than two hours the fire burned out of control. Collapsing walls made the work of firefighters extremely hazardous.

Again a general alarm was sounded, summoning all available equipment in the city. Airborne sparks from the blaze ignited secondary roof fires in other parts of the Downtown, and a gust of wind blew down the Pine Street wall, sending bricks crashing through the windows of the Providence Opera House across the street. Losses from the fire were estimated by the

Masons at $250,000. A year later the group rebuilt on the same site. The second Masonic building, still standing, has been recently rehabilitated for offices and retail use. Photograph of Union Depot (top), 1896, courtesy of Rhode Island Historical Society, Rhi x3 4535. Photograph of Masonic Temple (bottom), 1896, courtesy of the Providence Journal.

The destructive fires in the Downtown during 1896, coupled with the rapid growth of the permanent department, made the construction of a new central fire station a necessity. Accordingly, on June 13, 1900, the Rhode Island General Assembly authorized the Providence City Council to issue notes totaling $200,000 for capital improvements to the fire department. Several months later Gilbane Brothers began work on the new Central Fire Station. Designed by the architectural firm of Martin and Hall, the three-story brick and limestone building, complete with bell tower, became the new home of Hose Company No. 1, Ladder Company No. 1, and the administrative offices for the chief and the fire alarm service.

This photograph, taken on November 1, 1901, shows the building under construction. The Central Fire Station was occupied in March 1903 and remained headquarters for the Providence Fire Department until April 1935, when the property was sold to the federal government for a new post office building. It was demolished in 1938. Photograph, 1901, courtesy of Rhode Island Historical Society, Rhi x3 4947.

chapter THREE

The Era of Innovation, 1900-1951

As the twentieth century dawned, the sight of a three-hitch truck and a two-horse steamer charging to a blaze over cobblestoned main streets or churning up dust on unpaved side roads was still a common yet thrilling occurrence to the residents of Providence. Approximately one hundred well-groomed steeds from midwestern horse farms were under the care of Dr. Alexander C. Buchanan, department veterinarian for more than thirty years until his death in 1903. The education they received from George Hunt, director of the "Providence Training School for Fire Horses" at Engine 10, made their exploits and, indeed, their very presence a source of awe and amusement for young and old alike.

But the first half of the present century was an era of innovation. Horsepower gave way to the internal combustion engine, and steam power yielded to high-pressure water mains, rendering the steamer obsolete. Administratively, a pension program was established and the merit system was introduced, first for promotions and then for hiring. A two-platoon force was installed and drill and training schools were opened. The department inaugurated formal fire prevention programs, built functional modern stations, and created a rescue squad. The two-way radio revolutionized communications, while other technological advances rendered firefighting more effective. The mayor secured control of the department from the legislative branch of city government; then the union launched its successful attempt to organize the rank and file, and the chief gained control of appointments and promotions. In sum, during this half century the fire service assumed its present form.

The century began auspiciously for the department. On June 13, 1900, the General Assembly authorized the city to float a $200,000 loan for the construction of the new Central Fire Station to replace the already obsolete building that housed the "Three Ones." The new structure, built on a tract of land at the north border of Exchange Place, was a three-story brick and limestone building with a bell tower. The Board of Fire Commissioners, while proud of its new project, lamented in its 1901 report that a shortage of funds had prevented the construction of a fourth floor to house a gymnasium and school of instruction where "the members of the force would not only have been drilled in the duties pertaining to this service, but would also have had the opportunity of being exercised in a proper and systematic manner." The new headquarters was completed and occupied in March 1903. It was the centerpiece of a station-building boom that also produced new facilities on Point Street and Laurel Hill, Douglas, Mount Pleasant, and Humboldt avenues in the years from 1900 to 1907.

In 1901 the department registered two notable firsts. The Board of Fire Commissioners announced the establishment of competitive merit examinations for promotions. While the plan, said the commissioners in their annual report, "may appear to be a radical departure from former methods," it will benefit the department, for there is "no guarantee that because a man had served for a number of years that he is qualified, for that reason alone, to have command of men."

In July 1901 the board approved the pensioned retirement of Francis D. Chester of Engine Company No. 10 (Oxford Street) at the rate of six hundred dollars per year. This action made Chester—who had served the department for forty-five years—the first Providence fireman to be retired on pension.

As always, the fire service had its dangerous and tragic side. In January 1901 Captain Hiram D. Butts, a thirty-five-year veteran of the force, was thrown from a hose wagon drawn by a pair of untrained, or "green," horses and died of a fractured skull. Joseph Devine of Hose 13 was killed fighting a Dorrance Street fire in July 1902, and one year later, on July 27, 1903, twenty-six-year veteran John E. Carlin fell from a hose wagon while responding to an alarm. The wheels of his apparatus crushed his chest, causing instant death. In 1904 the department set a record for total alarms (1,066, including 8 third alarms and 2 general alarms) as fire losses reached $796,264, eclipsing the mark established in 1888. Included among the three-alarm blazes were fires at St. Vincent dePaul's Infant Asylum and St. Aloysius Home.

The rash of departmental fire deaths continued on January 24, 1907, when Captain George H. Noon succumbed to injuries caused by a fall that occurred while he was battling a South Water Street coal company blaze. On the day after Christmas, 1908, fifty-one-year-old ladderman Benjamin N. Brown died of a broken neck sustained in a fall while combatting a fire at an Ashburton Street coffin factory. With his death Brown became the tenth firefighter to give his life in the line of duty since the first fatality occurred in 1828.

In terms of lives lost, the city's most disastrous fire to

This is a typical Providence Fire Department three-horse hitch pulling a combination (hose-chemical) wagon around the turn of the century. Steamers were hauled by two larger horses. Prior to the introduction of motorized vehicles in 1911, the horse fleet numbered more than one hundred. Photograph, circa 1900, courtesy of John Ward.

This series of early twentieth-century postcards depicts the city's fire stations and apparatus then in use. The entire set, consisting of thirty-five postcards, is a valuable collector's item. Postcard views from the Patrick T. Conley collection.

a. Pictured (left) in front of the What Cheer Engine Company 15 house on Wickenden Street is a hose wagon built in 1892. It was equipped with rubber tires and carried eight hundred feet of 2½-inch hose.

b. Hose 20 on Manton Avenue used a combination wagon built in 1899 and carried 2½-inch as well as ¾-inch hose.

Oliver E. Greene Engine Co. № 18 House cor.
Broad and Rugby Sts., Providence, R.I.

c. The Oliver Greene Engine House (left) on
the corner of Broad and Rugby streets housed
an 8,500-pound second-size Metropolitan
piston-driven steamer that was placed in service
in 1906. Also pictured is a steel-frame combination wagon.

d. The Seagrave Trussed Truck pictured right
at the Ladder 8 station on Laurel Hill Avenue
contained ladders with an extention of fifty-five
feet. This ladder company consisted of a captain,
a driver, a lieutenant, and four laddermen.

Hook and Ladder Co. № 8 House Cor. of Laban St.
and Laurel Hill Ave., Providence, R.I.

Putnam Engine Co. № 14 House Cor.
Putnam and Amherst St., Providence, R.I.

e. Engine Company 14 on Putnam and Amherst Street received its third-size La France
piston engine in 1895.

67

f. The steamer at the Engine 16 station near the corner of Branch Avenue and Charles Street was the second of two second-size Metropolitan units purchased by the department in 1906.

90 Engine Co. Nº 16. Junction Branch Ave. and Charles St., Providence, R.I.

New Fire Station, Point Street, Providence, R.I.

g. The Point Street station was opened in 1908 and housed a ladder company (pictured above) as well as a first-size Metropolitan, built by the American-La France Fire Engine Company of Elmira, New York, and a combination wagon built by J. G. McIntosh and Son of Providence. The station had a gym on the third floor that in later days served as a center for departmental boxing and wrestling contests. This landmark station house of Engine No. 22 was demolished in 1971.

that date occurred on February 12, 1908, at the starch works of Charles S. Tanner on South Water Street. Five people were killed by an explosion that leveled Tanner's four-story brick building. But the fire that hit home for the department was the one at the Revere Rubber Company on May 7, 1912. That blaze claimed the lives of two firefighters, Lieutenant Christopher Carpenter and hoseman Harry H. Howe.

Such disasters, either actual or anticipated, prompted the passage of several state and local laws aimed at fire prevention. In addition to a rubbish inspection statute in 1898, the General Assembly in 1903 required the stationing of a fireman in a theater at any time when an audience was present. In August 1905 the City Council, responding to a detailed report on Providence prepared by the National Board of Fire Underwriters, passed an ordinance authorizing the inspection of all places where explosives were kept and requiring depart-

mental approval for storing "more than five gallons of naphtha, gasoline, benzine or any other product of petroleum which flashes or inflames at a less temperature than 110 degrees Fahrenheit" or garaging "more than three vehicles using any such product for motive power." This ordinance recognized the advent of the automobile and the gas-driven internal combustion engine—advances that would shortly revolutionize firefighting methods.

The first mention of the automobile in the department's annual reports came as early as 1899 under that section devoted to the horse fleet. In 1907 the board expressed "regret" that it had been "unable to secure the adoption of the automobile as part of the fire department equipment." Citing the experience of other cities, it recommended the purchase of autos for the use of "the chief and his assistants" so that superior officers could reach "the scene of fires much more quickly than under the present system." For the time being, however, the

Fire historian Franklin C. Clark refers to 1904 as the "Great Fire Year." During that twelve-month period Providence firemen responded to 1,066 alarms—a record to that time. Losses to property approached $800,000. On April 22 a chimney spark ignited the roof of the American Ship Windlass Company at the corner of Waterman and East River streets. The company was under government contract to produce machinery for several U.S. battleships then under construction that were destined to become part of Teddy Roosevelt's "Great White Fleet." The three-alarm blaze was fought by twenty-one pieces of fire equipment, including these steamers (top). The fire caused more than $100,000 in damage and delayed completion of the navy contracts.

Eight days later another three-alarm fire gutted the Anthony and Cowell furniture showrooms on Weybosset Street near Mathewson. The department's fire tower was called into action to douse flames on the upper floors of the six-story building. Newspapers of the day estimated that a crowd of twenty-five thousand watched as firemen rescued several employees trapped on the top floors. Estimates of the loss here approached $330,000.

In September firemen were put to the task once again as fires destroyed four city lumber yards. The year ended with a general-alarm fire at H. A. Grimwood's lumber yard on Westminster Street. Photograph, top, April 22, 1904, courtesy of Rhode Island Historical Society, Rhi x3 4952. Photograph, April 30, 1904, courtesy of Rhode Island Historical Society, Rhi x3 4096.

JOHN CUNNINGHAM, V. S.

LIKE RICHARD III, he's often heard,
This skillful veterinary,
In long discourse, to praise the horse
With learned commentary.
From splints to heaves, his care relieves
All equine ailments serious,
And this V. S., we must confess,
Has horse sense most mysterious.

Horses played an important role in the Providence fire service for almost seven decades. In 1859 rented horses were used to haul the heavy steamers. The first department-authorized purchase of horses took place in 1867, when the horse team pulling Engine No. 4 gained the distinction of being the first team so acquired.

Three years earlier a hostler, or groom, had been assigned to each steam-engine company. By 1870 the growing number of horses in the department necessitated the services of a veterinarian. For more than three decades Dr. Alexander C. Buchanan performed his task with distinction. After Buchanan's death in 1903, John T. Cunningham (above) continued the standard of excellence.

Besides ensuring the health and well-being of the equines under their care, the department's veterinarians often supervised the purchase of new animals for fire service. Most Providence horses were acquired from stables in Indiana, Ohio, or Michigan. The horses varied in weight from twelve hundred to fifteen hundred pounds, depending on the weight of the apparatus being pulled. During the later period a training school was established at Engine 10 on Oxford Street in South Providence. Generally two weeks were required to train horses for fire service.

At its height in 1911 the Providence Fire Department stabled 120 horses. Nine years later there were none. Line drawing by Farnum, courtesy of Rhode Island Historical Society, Rhi x3 4948.

As important as any chief of the fire department was Ira Winsor. His period of service spanned fifty-two years (1857-1909), not counting his time as a youthful torchbearer in the volunteer era. From his first assignment in Hand Engine Company No. 1, he moved to first hoseman in Steam Fire Engine Company No. 1 in 1860 and attained the rank of first assistant engineer in 1875, a post he held until retiring as an active fireman in 1883. In 1885 he entered the Common Council and ten years later became an alderman. His tenure on the City Council continued until 1902, and during all of his seventeen-year incumbency he served as either member or chairman of the Committee on the Fire Department. Winsor also doubled as fire marshal from March 1888 to January 1904. He capped his distinguished career by presiding over the department as a member and chairman of the Board of Fire Commissioners from March 1902 until a well-deserved retirement in January 1909. Engraved portrait by Ryder-Dearth, courtesy of Rhode Island Historical Society, Rhi x3 5170.

The central fire alarm was moved from City Hall to the third floor of the new Central Fire Station in 1905. Two operators manned the alarm room at all times. When a box alarm was pulled, a red light automatically flashed in the central station on the circuit connected to the box and a small gong was activated that sounded the number of the box. A pen register then recorded the number on a broad strip of paper. At the first stroke of the gong, the operator pulled a small lever that electronically alerted every station house in the city. The station alarm turned on all lights, and it opened stall doors to allow horses to position themselves quickly for harnessing. With the entire department in readiness, the operator then sent out two rounds of the number of the box pulled, prompting action by the appropriate hose and engine company. All other firemen returned to their quarters. Providence was one of the few large cities to alert the entire department before specifically designating the location of a fire. Photograph, 1905, courtesy of Rhode Island Historical Society, Rhi x3 4948.

department contented itself with the purchase of a new Seagrave Aerial Hook and Ladder Truck, two new combination wagons built locally by John G. McIntosh and Son, and two steam engines, one of which, a famed Amoskeag, was now being manufactured by the International Power Company of Providence.

During this century's first decade several important changes in fire personnel occurred. In 1902 Stillman White, first chairman of the Board of Fire Commissioners, resigned. His spot on the department's governing body was taken by fire marshal Ira Winsor (1888-1904), who had advanced through the ranks of the fire service and then had helped to direct fire policy as a member of the City Council and its Standing Committee on the Fire Department. Winsor remained a commissioner until 1908, capping a long career of distinguished fire service.

In 1909 George Steere retired, having served as chief since 1884, concluding the longest tenure of any Providence fire chief before or since. Holden Hill, Steere's efficient and compatible deputy since 1885, was rewarded with the position of chief for three months prior to his own retirement after forty-six years on the force. On July 1, 1909, district chief Reuben D. Weekes of Station 12 (Smith and Orms streets) assumed direction of the ever-expanding department.

As the retirement bell tolled for Steere and Hill, the clanging of the big outside alarm bells ceased forever. The system of bell-ringing, begun first from the steeples of churches during the volunteer era to mobilize the townspeople and continued from the towers of fire stations to alert the call men, was discontinued in January 1909. The commissioners justified this abandonment of tradition by citing the existence of a completely professional force. The bells no longer serve a useful purpose, said the board, but merely "complicate matters by collecting crowds which bother the police and interfere with the firemen."

At the changing of the guard in 1909, the *Providence Board of Trade Journal* published a feature article that praised the virtues of a fire force whose ranks had increased to 330 men organized into thirty-three companies—an enlargement due in part to increased duties created by the recent fire inspection laws. The article described a department "admirably equipped with modern and reliable apparatus," all horse-drawn, including 17 steamers,

13 combination wagons with chemical tanks, 11 hose wagons, 11 hook-and-ladder trucks, 2 protective wagons, and 1 sixty-five-foot water tower. The *Journal* was particularly impressed with the Seagrave ladder. Whereas the 1883 Skinner truck required six minutes and eight men to elevate it, the new Seagrave, operated by the same company (Hook and Ladder No. 1), could be raised to a comparable seventy-five-foot height by two men in only fourteen seconds.

During the second decade of the century the department experienced dramatic modernization under the progressive direction of Chief Weekes and Board Chairman John R. Dennis, the former fire marshal. Far and away the most significant innovation was motorization. Like the changeover from manpower to horsepower, this revolution in locomotion was accomplished gradually. The motor vehicle was introduced in January 1911, when the insurance interests bought a salvage wagon for the Protective Company, a unit which, ironically, was dissolved in January 1913 when the insurance interests and the City Council disagreed over its maintenance. The first city-owned machine, a Knox combination truck, was placed in service at the Putnam Street fire station on May 25, 1911. For nearly a decade the process of replacement continued. Horse-drawn trucks were discarded in favor of "power apparatus," while simultaneously the colorful steamers succumbed to high-pressure hydrants and triple combination trucks. In December 1920 the last piece of horse-drawn apparatus was decommissioned, and at year's end the all-motorized department consisted of fourteen triple combinations (pumper, hose, and chemical), twelve double combinations (hose and chemical), twelve hook-and-ladder trucks, and several auxiliary vehicles for the use of department officers. With the horses out to pasture (or in reserve), yet another memorable era came to a close.

Chief Weekes relished his job as chief and performed it with a flair for public relations. He authored a brief survey of the department for a magazine entitled *American City*, developed fire prevention programs in concert with the Chamber of Commerce, and hosted the 1916 convention of the International Association of Fire Engineers, whose second and last visit to Providence attracted approximately fifteen hundred participants. Features of

The Central Street fire station near Hoyle Square opened in 1875 and served the West End neighborhood for sixty years. This 1906 photograph shows the members of Good Will Engine Company No. 13 with their new first-size La France piston steam engine. This steamer was drawn by three horses and weighed more than five tons. The combination wagon was originally built as a hose wagon in 1892 by Lewis Falls of Providence. Pictured standing at left in front of the combination wagon are Captain William Willard and Lieutenant Owen Trainor. The mascot seated on the combination is not identified. Photograph, courtesy of Rhode Island Historical Society, Rhi x3 1173.

Although Holden O. Hill served as chief of the Providence Fire Department for only three months (April 1 to July 1, 1909), his service with the department spanned more than four decades. Hill was born in Foster, Rhode Island, in 1840 and came to Providence just after the outbreak of the Civil War. He joined Atlantic Hand Engine Company No. 10 in June 1863 and four years later became a hoseman with Engine 8 on Harrison Street. In 1883 Hill was elected assistant engineer, two years later he became the city's second deputy chief engineer when George Steere left that post to head the department. Hill served as deputy until 1909, when, at midnight on April 1, he became head of the department while directing a company of firemen at a small blaze on Sabin Street. After retiring from the fire service on July 1, 1909, he remained active as a member of the Providence Veteran Firemen's Association. Photograph, courtesy of Rhode Island Historical Society, Rhi x3 5153.

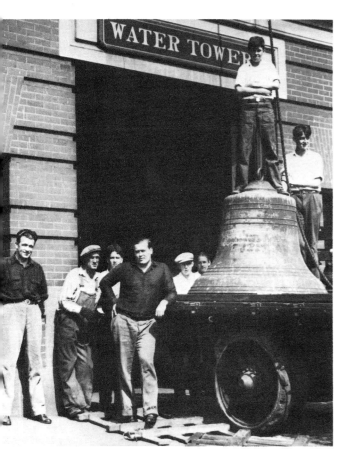

The establishment of the central alarm system made the bell tower alarms increasingly obsolete. At midnight on June 1, 1909, the three-thousand-pound bell at the Central Fire Station rang for the last time. It was taken down in 1931, placed on the truck shown here, and carted off for junk.

The old bell once operated by Gardner T. Swarts near the corner of Pine and Richmond streets fared better. It had been transferred to the "Seven's" fire station on Richmond Street. It was taken from that station sometime during 1909, stored for two years, then sold to the city of Cranston, which installed it in the tower of the Eden Park School House. Photograph, 1931, courtesy of the Providence Journal.

that large and successful August convention included the creation of a professionally directed twelve-man musical chorus of firefighters to serenade the visitors and a giant exhibit of firefighting apparatus at the Cranston Street Armory.

The first two decades of the twentieth century, called by historians "the Progressive Era," were years of sweeping municipal reform—most of which bypassed Providence. One cause that the local leaders did embrace, however, was housing and building-code reform. In April 1908 the General Assembly passed "An Act to Diminish Danger to Life in Case of Fire," requiring all doors and exits in theaters, schools, halls, factories, hospitals, asylums, and churches to open outward and requiring every theater to have an automatic sprinkler system, a central ventilator, and a fire alarm box. The fire department did the surveillance and the local building inspector enforced the act. In 1911 renovations in City Hall were necessitated by a statute that required the fireproofing of all public records.

The catalyst for more significant reform was a comprehensive survey of residential structures made by John Ihlder and his associates and published in 1916. Ihlder's report, entitled *The Houses of Providence*, described the marked increase in the construction of multiple-family dwellings in the years since 1900, when single-family houses outnumbered all others by a ratio of 3 to 2. During the five years from 1911 through 1915, said Ihlder, "more families have been provided for in three-deckers than in one- and two-family houses combined." Such a building trend produced congested districts, and these wood-

framed houses posed a serious "fire hazard" that was causing "increased expense for insurance, fire apparatus, hydrants and water supply." The report's major recommendation was a "thoroughgoing housing code" which would prevent land overcrowding on tiny lots and require the walls and roofs of buildings crowded closely together to be made of nonflammable materials. The study also urged sanitation improvements and a zoning scheme.

As might have been expected, it took a catastrophe to bring about serious consideration of the proposals of Ihlder and other preventionists. The tragedy occurred on the evening of January 31, 1921, when the brick walls of the Washington Bowling Alley, a nonfireproofed structure on Mathewson Street, crashed down on two dozen firefighters, killing four (Thomas H. Kelleher, John J. Tague, Arthur Cooper, and Lieutenant Michael J. Kiernan) and critically injuring about twenty others. At the urging of Mayor Joseph Gainer, the City Council doubled the firemen's relief fund to allow appropriations of ten thousand dollars to the dependents of each man killed in the line of duty, and it also proposed substantial revisions in the building code, with particular attention to fireproofing all Downtown structures. Unfortunately, owners of existing buildings resisted and delayed passage of these safety requirements because of the expense such changes would cause them to incur. In June 1923 Providence passed its first zoning ordinance, but not until January 1, 1927, did the council impose a comprehensive building code embracing some of Ihlder's safety suggestions.

During this era the National Fire Protection Association gained public acceptance of its concept for a national fire prevention observance. On October 9, 1916—the forty-fifth anniversary of the Great Chicago Fire—the first fire prevention day was declared, as Mayor Gainer and Governor A. Livingston Beekman, in step with many public officials throughout the country, issued proclamations. By 1922 the day had become a week, with the Providence Safety Council and the Fire Prevention Committee of the Chamber of Commerce spearheading the observance and exhorting the public to take "the very best precautions to avoid unnecessary losses by fire." Many groups got behind this pioneering effort. The Boy Scouts distributed posters, the clergy addressed it in their Sunday sermons of October 1 and 8, the school department held fire drills and essay competitions, and the uniformed members of the fire department distributed twenty thousand circulars to schools and organizations. The new medium of commercial radio was also called into play, with the fire prevention message being broadcast from the Shepard Company's recently established station (WEAN) by Battalion Chief Frank Charlesworth and over the Outlet Company's station (WJAR) by Fire Prevention Committee Chairman Henry A. Fiske. Such public support led Mayor Gainer, Charlesworth, and the chamber to secure the creation in 1925 of a Fire Prevention Bureau as a branch of the department.

In 1921 Reuben Weekes retired. For the next three decades, despite the department's undiminished efficiency and leadership, the history of the Providence fire service becomes difficult to relate. From the creation of the paid force in 1854 until 1908, the department (like all other municipal agencies in Providence) published detailed annual reports. Early in the Weekes regime, several departmental updates appeared in the *Board of Trade Journal*, a businessman's weekly that continually extolled the excellence of Providence's fire service. After 1916, however, the only sources for department history are scattered news stories in the *Providence Journal*, the sparse and random minutes of the City Council's Committee on the Fire Department, and the brief, stark statistical listings in the yearly *Pocket Manual* of the council, the *Providence City Directory*, and the *Journal Almanac*.

Weekes was succeeded in August 1921 by Third Ballation leader William F. Smith, then a thirty-five-year

veteran of the force. During his short year-and-a-half tenure, Smith put his stamp on the department by successfully lobbying for the long overdue creation of a two-platoon system. Under the old system put into effect in 1899, the firefighter had one day off in five, but during the other four he was on duty continuously. Under the initial and complicated two-platoon plan, he was allowed one day off in six, but his hours of actual duty were greatly reduced. Men were to work in two shifts—days of ten hours' duration or nights of fourteen—with each man working three turns on the day shift and then an equal number of nights. Each changeover gave one-third of the department (or battalion) its free day.

In February 1923 the City Council rejected the scheme because of its cost: the approximately 100 additional men required would increase the department's budget by nearly $250,000 and bring the total number of firefighters to 420. In March, however, the council had a change of heart after amending the ordinance to prohibit members of the department from "pursuing other gainful occupations while employed by the city." A second administrative requirement also detracted from this reform. Under the new system no man, either on his full day or his partial day off, was allowed to leave the city without the "special permission" of the chief.

The two-platoon system at last provided some measure of leisure time for the Providence firefighter. Since such leisure coincided with the advent of the Golden Age of American Sports, firemen turned increasingly to organized athletic competition. Intradepartmental contests were held in baseball, wrestling, and boxing during the twenties and thirties, with the best in the fire service squaring off against the police department champions. The gymnasium in the Point Street station served as the training center for many an off-duty grappler or pugilist. Although most of these competitors and their exploits have been obscured by time, the feats of some endure—like those of Jim Killilea, boxing and wrestling champion of the department which he later headed, and boxing titlist Frank Moise, father of the present chief.

With Providence firefighters engaged in more currently popular athletic pastimes, muster competitions in these years were dominated by volunteer firemen from the rural towns, especially East Greenwich, East Provi

Reuben DeM. Weekes began his fire department career in November 1882 as a hoseman for Hose Company No. 4. After fighting such blazes as the Aldrich House and Theatre Comique disasters of 1888, he was given a position as assistant foreman with the Water Witch Hose Company on Benevolent Street. Then he returned to Hose 4 on Mill Street as a lieutenant. In 1906 he was made district chief, and on July 1, 1909, he became chief of the department.

Shortly after assuming his new duties, Weekes began motorizing the fire service—a process that began in 1911 and concluded on December 9, 1920, when the last piece of horse-drawn apparatus was withdrawn from service.

Weekes's improvement program also resulted in the replacement of the department's steamers by motorized pumpers. The Providence chief worked closely with the Chamber of Commerce to remove the city's fire alarm signal station from the top floor of the Central Fire Station to a new fireproof building on Kinsley Avenue by 1919.

When Weekes left office after forty-one years of service, the department consisted of thirty-three fire companies and a permanent force of more than 350 men. In October 1945 Weekes died after a long illness at the age of eighty-six. Photograph, courtesy of Rhode Island Historical Society, Rhi x3 4954.

The first piece of motor-driven fire apparatus was purchased for the Providence Protective Company in 1911. This nine-member force had been organized in 1875 to "protect and save as far as possible life and property in case of fire in the city." It grew out of the concern that in many cases water caused far more damage to property than the fire itself. Expenses for the upkeep of this company were shared by the city's

Board of Underwriters.

The first protective wagon was placed in the Three Ones station on Exchange Place, but in April 1891 the company moved to Richmond Street. The forty-eight-horsepower truck pictured here was purchased from the La France Company of Elmira, New York.

Ironically, only a year after the addition of this new piece of equipment, the insurance

companies decided to **withdraw their financial** support for the unit. After a bitter debate in the City Council, the Providence Protective Company passed into history at midnight on January 11, 1913. The La France truck was sold and served the Richmond, Rhode Island, volunteer fire department until the late 1930s. Photograph, circa 1911, courtesy of Rhode Island Historical Society.

The Niagara Company No. 1 of New London, Connecticut, is generally credited with operating the nation's first motor-driven fire department vehicle in 1903. As early as 1899 the Providence Board of Fire Commissioners suggested "that the instant the practicability of the automobile, or other similar motor, for fire department purposes is assessed, it should be made a part of our equipment, as its introduction would mean the saving of thousands of dollars over the present cost of maintenance." Again in 1906 the commissioners restated the need to test

motorized equipment for the Providence force.

It was not until 1911—only a few months after the introduction of the Protective Company's truck to the Providence fire service—that city authorization was given for the purchase of the fire department's first motorized vehicle. On May 25, 1911, Engine 14 on Putnam Street replaced its horse-drawn hose wagon with a combination truck built by the Knox Automobile Company of Springfield, Massachusetts. Charles J. Prendergast (pictured at the wheel) became the department's first

"chauffeur." The Knox car was replaced by a triple combination wagon in 1920. Ironically, less than six months after Engine 14 made its first run, a lengthy article in the Providence Journal confidently claimed that the horse-drawn fire apparatus was in "no immediate danger" of being phased out. "Perhaps in the dim and distant future," said the Journal, "the fire horses will become obsolete." That "dim and distant future" proved to be less than a decade away. Photograph, 1911, from the Patrick T. Conley collection.

With City Council support, the department's phaseout of horse-drawn apparatus accelerated. In February 1913 two more Knox combination trucks were purchased and assigned to Hose 5 on Hope Street and Hose 11 on Elmwood Avenue. The last horse-drawn vehicles made their final run at 2:36 P.M. on December 1, 1920. In a symbolic gesture, Mayor Joseph H. Gainer at that moment pulled Box 146 at the corner of North Main and Market Square. The alarm brought the city's two remaining pieces of horse-led equipment—Engine No. 2's hose wagon and the combination wagon from the "Seven's" on Richmond Street. Both companies arrived at the scene expecting a fire, but they found instead a host of city officials and two new gasoline-powered combination wagons. The two teams of horses were sold a short time later, ending one of the most colorful periods in the history of the department.

Elsewhere, New York's fire department was completely motorized by 1922. In Philadelphia the tradition of horse-drawn equipment lingered on until New Year's Eve of 1927. Photograph, 1913, courtesy of the Providence Journal.

The fire department suffered a double fatality on the afternoon of May 7, 1912. At about 2:00 P.M. fire broke out at the storage warehouse of the Revere Rubber Company on Atwells Avenue. Two workers trapped on the fourth floor smashed windows and leaped onto the roof of the William Harris Company office next door. By the time firemen arrived, the building was completely involved. Rubber products stored in the warehouse produced intense heat and enveloped the area with a dense, pungent shroud of smoke.

Police forced crowds back when it was learned that a large tank of gasoline was in the burning building and might blow up at any moment. Hoseman Harry H. Howe and Lieutenant William Cameron were atop the Ladder 6 aerial when an explosion hurled them to the ground. Cameron survived the fall, but Howe received serious head injuries and died several days later. Meanwhile, Lieutenant Christopher Carpenter of Ladder 6 was also fatally injured in a fall while battling the blaze from the roof of the adjacent warehouse.

Loss to the Revere Company was estimated at more than a half million dollars. The fire department's loss could not be measured in material terms, and its effects would remain long after the ashes were cleared from the Atwells Avenue site. Photograph, May 7, 1912, courtesy of the Providence Journal.

dence, and Westerly. Some of these volunteers manned Providence hand-pumpers from an earlier age. The conspicuous exception was the Nonantum, an engine sponsored by the Providence Veteran Firemen's Association, which won a statewide competition as late as August 1932. Regional hand-engine musters, first held on July 4, 1849, in Bath, Maine, were faithfully reported on the sports pages by the local press until World War II interrupted these historic competitions.

Battalion Chief Frank Charlesworth, an English immigrant, assumed the leadership of the department on March 10, 1923, when William Smith, fresh from his two-platoon victory, retired. Charlesworth also pressed for administrative changes. Upon his return from the 1927 national fire chiefs' convention, he issued a detailed statement to the fire commissioners and the City Council recommending the establishment of a training school complete with tower, smoke house, and other firefighting aids. Lamenting Providence's lack, Charlesworth asserted that "practically every city of real consequence in the country has had a training school for years." The chief could have added that such a facility had been recommended by the Providence Board of Fire Commissioners as far back as its annual report for 1907.

The response to Charlesworth's appeal, while not instantaneous, was prompt. In December 1928 construction began on a combination drill tower-smoke house on Whitmarsh Street adjacent to the Bucklin ball field. Here the rookie would be schooled in jumping, hose handling, and gas mask use. On land north of the tower a paint and repair shop was begun, with classroom space therein where "the theory of fighting fire is to be expounded by a veteran of the calling," as one news account described this pioneering educational effort. Other new pre-Depression construction included fire stations on Academy Avenue (1927) and Rochambeau Avenue (1929).

In 1931, upon the recommendation of Charlesworth, Mayor James Dunne, and the Chamber of Commerce, a municipally maintained salvage corps was established, with duties similar to those of the old Protective Company which had been disbanded in 1913 when the city and the insurance companies failed to agree on the financing of the unit. According to the chamber's recommendation,

"the salvage corps would aid greatly in reducing incidental loss through prompt action in caring for the contents of burning buildings."

In the same year the decade-long controversy over the purchase of a fireboat was resolved by a very modest concession to its advocates. The City Council authorized the expenditure of one thousand dollars to put some rudimentary firefighting apparatus on the harbormaster's boat in the aftermath of the great fire that destroyed State Pier No. 1. New York we were not.

Apart from these small ventures, little growth or building occurred during the early years of the Great Depression as Providence, along with the rest of the nation, plunged into debt. The first new piece of apparatus acquired during the decade of the 1930s—an eighty-five-foot aerial ladder truck—was not purchased until October 1937.

Although the department was stymied in its efforts to modernize, firefighters' jobs were secure. When the national unemployment rate topped 25 percent by early 1933, the ranks of the fire service remained full. In 1930, the year following the Great Crash, total personnel—including officers and clerks—was 489; in 1933 it was also 489, and it remained constant throughout the Depression decade. In fact, the Providence Governmental Research Bureau, in a 1934 survey of the department, concluded that the city had "more stations and companies than it needs." This surplus, said the bureau, "has been the result of substituting motor-driven apparatus for horse-drawn apparatus without making due allowance for the increased area the motor-driven apparatus is capable of covering."

In the gloomy 1930s the most spectacular activity affecting the fire service was the political fireworks between once dominant Republicans and insurgent Democrats for control of departmental patronage and policy. To be understood, this struggle must be viewed in its historical and constitutional setting. Both the paid force and the Republican party were founded in 1854. Within two years of its birth the GOP had seized control of the city government, which was operating under a charter that gave most of the power and patronage to the bicameral City Council and made the mayor institutionally impotent. This strong council-weak mayor system prevailed until a new city charter went into effect in January 1941, so into the decade of the thirties the party that controlled the council controlled the fire department. Until 1928 that party was Republican.

The manner in which the Republicans had managed to maintain political supremacy in Providence despite the numerical edge enjoyed by Irish Democrats is evidence of GOP resourcefulness. By 1888 the real estate requirement for voting and officeholding imposed upon naturalized citizens by the state constitution no longer served its original purpose—that of depriving Irish-born Catholic Democrats of political power. Thus an incongruous coalition of WASP Republicans and reformers secured its removal. The reform (Article of Amendment VIII of the state constitution) was also a politically inspired maneuver, limiting the vote in City Council elections to those who paid a property tax. A 1925 study indicated that of the Providence electorate who voted for mayor, nearly 60 percent—principally ethnics at the lower socioeconomic levels—were constitutionally ineligible to cast a ballot for their councilman. The importance of this voting restriction becomes evident when one remembers the relative governmental strength of the council and the mayor. From 1896 onward the ceremonial chief executive (for whom every voter could cast a ballot) was nearly always a Democrat, but the two-chamber council, controlled by Yankee Republicans, in turn controlled the city finances and most of the patronage, including appointments to the Board of Fire Commissioners. Small wonder that the Democrats battled ceaselessly to remove the property tax requirement. After they finally succeeded in 1928 (via the Twentieth Amendment), the council came permanently under Democratic control, an event which at last allowed the buildup of an effective political machine.

But the Republicans, who still controlled the rural-dominated General Assembly, refused to yield their long-held power without a vigorous fight. In 1930 the state legislature amended the city's charter, reducing the number of ward councilmen from four to three and increasing the number of wards from ten to thirteen. These changes were a bold but unsuccessful effort by the endangered GOP to keep control of the council by gerrymandering it.

The development of the modern skyscraper during the last decade of the nineteenth century brought with it new challenges for metropolitan fire departments. In 1879 the water tower was introduced in fighting high-rise fires in New York. Soon water towers became an essential piece of fire apparatus in most large American cities. Providence purchased its Champion water tower in 1893. Built by Chicago's Fire Extinguisher Manufacturing Company, the tower telescoped by hydraulic pressure to a height of sixty-five feet. The tower's nozzle was controlled by two ropes held by firefighters on the ground. In 1914 the water tower was fitted with a new adjustable metal pipe replacing the old rubber pipe. In a test (shown here) conducted on Gaspee Street near the old Brown and Sharpe plant, the tower, aided by a steamer pumping water through three hose lines, shot a stream of water 150 feet across the Woonasquatucket River. With its two-inch deck nozzle, the tower could throw two thousand gallons of water per minute. Photograph, courtesy of Rhode Island Historical Society, Rhi x3 4933.

The Providence Board of Trade, and its successor the Providence Chamber of Commerce, have been consistent and effective voices supporting the efforts of the Providence Fire Department. In 1914 the chamber's Civic Affairs Committee, chaired by John Hutchins Cady, produced a series of recommendations aimed at improving the quality of fire protection in the city. The committee's proposals, submitted to the city's Board of Fire Commissioners, included a recommendation that the fire alarm headquarters be removed from the top floor of a brick building with interior wood construction to a separate building, "absolutely fire proof, where there could be no possibility of a fire or interruption of service."

During the next four years chamber members lobbied and appeared before numerous committees of the City Council, promoting their plan to remove the central fire alarm from the Central Fire Station. In August 1919 their hard work paid off when the new central fire alarm system took up headquarters in an unpreten-

tious two-story building on Kinsley Avenue (shown here). This new fifty thousand dollar building was touted as being completely fireproof, containing cement floors and steel doors, paneling, and window frames. Providence, in fact, was one of the first major cities to house its central alarm in a separate building, prompting Providence's superintendent of alarm, Gilbert S. Inman, to assert that the department had the best-equipped central alarm in the nation.

The new system contained a series of twenty-seven registers for each circuit, replacing the single register in operation at the central station. This change dramatically reduced the possibility of confusion when alarms came in simultaneously. These registers, such as the ones shown in this photo, punched a series of holes in a paper tape to indicate the alarm box number. Upon receiving an alarm, the operator set a dial for a "fast alarm" that was transmitted to the fire department. He then set a "slow dial," which fed a message to the police and newspaper offices. With fires in the Downtown, alarm opera-

tors turned another key that sounded horns posted at important intersections in the city, warning traffic police that department apparatus was coming in their direction. The entire system received power from 1,853 storage batteries housed on the alarm building's first floor.

In the years following the construction of this new departmental nerve center, two state laws increased its functions. In 1919 the General Assembly passed a statute mandating the installation, at owner's expense, of fire alarm boxes in every theater, and ten years later the same requirement was imposed upon schools, hospitals, asylums, and orphanages.

Presently a new $2.2 million communications center is under construction on West Exchange Street. This 1919 structure will be demolished to facilitate the development of the Capital Center Project. Photograph, top, 1919, from the Board of Trade Journal, April 1919. Photograph, August 24, 1919, courtesy of the Providence Journal.

The city's rapidly expanding immigrant population produced a corresponding growth in crowded, substandard housing. Concern over unhealthful conditions and the increased risk of a major fire prompted the formation of a General Committee on Improved Housing in 1915. John Ihlder, of New Rochelle, New York, was retained by the group to conduct a survey of existing housing in the city. A year later he issued a report, The Houses of Providence, that produced a series of recommendations designed to eliminate the conditions pictured in this top photo. Ihlder's major remedy was the enactment of a comprehensive housing code. Under a section entitled "Lessons Taught by Experience," Ihlder issued this prophetic warning: "While Providence is spared such a lesson as neighboring cities have had, its builders probably will continue to accept their risk, for the chance of mishap to each seems too remote to warrant much uneasiness." In 1917 housing-

code legislation was introduced in the General Assembly but was quickly killed because of the cost to property owners that such regulations would entail.

Another critic who expressed concern over this unsafe situation was Journal cartoonist Milton Halladay. As early as 1914 he graphically warned of the impending danger (cartoon, lower left).

The fears of Ihlder and Halladay were realized in the early morning hours of February 1, 1921. During a blaze at the Washington Bowling Alley on Mathewson Street, a wall collapsed, killing three firemen instantly and fatally injuring a fourth. Flimsy construction was blamed for the tragedy. Halladay's second cartoon poignantly demonstrates the failure of reform.

Public outrage, aided by the active support of local newspapers, eventually produced results. In April the state legislature passed an act au-

thorizing Providence and other cities to enact zoning ordinances. Within a year the Providence City Council began preparations for the passage of a zoning law. In April 1923 a draft of the ordinance, with a map depicting the city's proposed zoning districts, was presented to the City Council. Two months later the council approved the plan.

The General Assembly also passed an act allowing the capital city to enact a building code, to be supervised by a building board of review. The council responded slowly, but a comprehensive building ordinance finally went into effect on January 1, 1927. Photograph, 1916, courtesy of Rhode Island Historical Society, Rhi x3 4956. Cartoons by Milton Halladay, 1914 and 1921, courtesy of Rhode Island Historical Society, Rhi x3 4621 and 5151.

When this move failed to prevent Democratic victory in the 1930 elections, the state legislature abolished the council-controlled Board of Fire Commissioners in April 1931 and created the Board of Public Safety, a *state* agency to run the Providence fire and police departments, with power not only over policy but over patronage as well. Immediately the new governing body introduced physical fitness tests designed in part to retire some whom the Democrats had just appointed or promoted.

The council and Mayor James E. Dunne brought suit to enjoin what they considered a blatant infringement on the right of local self-government, but the Republican-dominated state Supreme Court upheld the statute in the landmark case of *Providence* v. *Benjamin P. Moulton, George T. Marsh, and Michael H. Corrigan.* The three successful defendants constituted the board that ran the fire service until a Democratic Assembly unceremoniously abolished it during the famed "Bloodless Revolution" of January 1, 1935. Governor Theodore Francis Green then appointed State Police Superintendent Edward J. Kelly to fill the administrative void that existed until June 1935, when the legislature created a Bureau of Police and Fire. This mayorally appointed three-man city agency was given "full control over, and management of the Police and Fire Departments." Its initial members were Benjamin Moulton, a highly regarded former member of the Republican Board of Public Safety, and two prominent Democrats—Chairman Thomas H. Roberts, who later became chief justice of the Rhode Island Supreme Court, and Joseph C. Scuncio. With the creation of the bureau two major innovations were made: the mayor gained a decisive role in fire department affairs, and the Democratic party got the chance to influence the personnel practices of the fire service.

During the 1930s the city began to acquire its national reputation for excellence in fire prevention. With Chief Charlesworth leading the way (and even staging his own public demonstrations of preventive techniques), Providence won several national awards, including first prize in the low fire-loss and fire prevention contest conducted by the United States Chamber of Commerce. In 1933, 1934, and 1936 the city ranked first in this field among municipalities in the 250,000-to-500,000 popula-

tion category. In 1934 it also won the grand prize over all other American cities. This effort had reduced Providence's fire loss by 1936 to 58.2 cents per capita, the lowest rate in the city's history.

Despite such highly acclaimed precautions, a few major fires marred the decade. In February 1931 flames destroyed State Pier No. 1 on Allens Avenue near the foot of Public Street, causing a loss approaching a half million dollars. The opening of this major city landmark in 1913 had made Providence a leading immigrant landing station. Here ships of the Fabre Line had carried thousands of immigrants—especially from Italy, mainland Portugal, and the Azores—to their new homes in America. The restrictions and discriminatory quota system established by Congress in 1924 had sharply reduced the flow of immigrants to the processing rooms on the structure's second level, and the devastating fire of 1931 practically ended Providence's role as a port of entry.

At the end of the decade the harbor was again the scene of major conflagrations. On July 6, 1939, a spectacular four-alarm blaze injured three firefighters, destroyed $100,000 worth of property at the Eastern Coal Company on Point Street, and endangered the Manchester Street plant of Narragansett Electric. According to newspaper accounts, the river was aglow for miles and the flames, leaping skyward, were observed from as far south as Warren. Three months later McCarthy's Freight Terminal on Allens Avenue went ablaze, producing damage also estimated at $100,000. The port of Providence was also ravaged by the winds and tidal wave of the 1938 hurricane, which completely flooded Downtown and placed an enormous cleanup and bail-out burden upon the fire service.

As a means of combatting the record unemployment spawned by the Great Depression, the federal government embarked upon a massive work relief program. As early as 1932 the U.S. Treasury Department invited proposals from the city for the sale or donation of land, convenient to the railroad, on which the federal government could erect a post office. Following Washington's rejection of several sites, the City Plan Commission proposed the area occupied in part by the Central Fire Station. After some City Council resistance because of the inconvenience such a move would cause the fire department, the administra-

William F. Smith's career with the Providence Fire Department spanned more than thirty-seven years. An admitted fire buff, Smith joined Engine 8 on Harrison Street as a call hoseman on November 1, 1886. Three years later he was appointed a permanent ladderman and moved to Ladder No. 1 on Exchange Place. In 1895 Smith became a captain, and on June 1, 1902, he was made a battalion chief of the Third Fire District and assigned to headquarters. He remained in that post for nearly two decades, working out of his headquarters at Station 22 on Point Street.

In August 1921 Smith was chosen to succeed Reuben Weekes as chief. Weekes and the Board of Fire Commissioners had tried unsuccessfully to convince the City Council to establish a two-platoon system in the department. Smith worked out a revised staffing plan that won the unanimous approval of the council on March 6, 1923—just four days before he stepped down as chief. Smith's successful fight to establish the two-platoon system raised the effective force from 295 men in 1921 to 420 men in 1923, an increase that put the annual budget of the fire department over the million dollar mark. Smith died on March 16, 1937, at the age of seventy-six. Photograph, 1916, courtesy of Rhode Island Historical Society, Rhi x3 5023.

The block-long Shepard's Department Store had been a Downtown landmark for more than a half century. Shortly before 8:30 on the night of March 8, 1923, fire broke out in a ventilating shaft that led from a stove in the kitchen behind the store's restaurant. Quick work by the fire department, combined with the apparently successful operation of the building's Grinnell sprinklers, seemed to extinguish the flash fire. A recall was sounded from Box 581 at 8:49, and several fire companies returned to their stations. Grinnell workers had arrived on the scene and shut off the water to make repairs on the sprinkler system. But several minutes later a second alarm was sounded from a box at the corner of Westminster and Snow streets—flames had traveled up the shaft and had set fire to the roof of the building.

By 9:45 nearly the entire department was on the scene. Seventeen-degree temperatures hampered the firefighting effort. Although the flames were confined to the fifth and sixth floors of the Washington Street end of the store, tons of water cascaded onto the floors below, ruining thousands of dollars in merchandise. Water was ankle-deep in the basement store. When it was all over, more than $1.5 million in damage had been inflicted and almost twelve hundred employees were temporarily out of work. Photograph, March 8, 1923, courtesy of Rhode Island Historical Society, Rhi x3 4936.

Frank Charlesworth, a native of England, began his career with the fire department in 1898 as a hose driver for Engine 14 on Putnam Street. His ambition and total dedication to the fire service were quickly recognized. In 1908 Charlesworth was promoted to the rank of lieutenant, and within a year he had won the captain's position at Hose No. 3 on Franklin Street. His rise through the ranks continued in 1917 with his selection as battalion chief. Charlesworth's appointment as chief of the department became effective on the night of March 8, 1923, while he was in the thick of the fight to control the blaze at the Shepard Company building.

Known affectionately as "Bucky," the new chief soon developed a reputation for the kind of aggressive leadership that was to gain the Providence department a national reputation. In 1928 he founded the drill and training school, and during the 1930s he initiated a tough inspection campaign directed at eliminating combustible materials from residential, industrial, and commercial buildings in the city. Results of the inspection program were immediate and dramatic. In 1933, 1934, and 1936 the depart-

ment won national fire prevention awards, and the United States Chamber of Commerce in 1936 cited Providence for having the lowest fire loss of any city of its size in the country. That year the fire loss was fifty-eight cents per capita—the lowest in the city's history.

As chief, Charlesworth remained a frontline firefighter. During his forty years with the force, he was twice cited for heroism in rescuing victims from tenement fires. Although injured during the 1931 blaze at the state pier, the chief remained at his position, directing the operations of the men under his command.

Governor Robert E. Quinn named Chief Charlesworth the state's first fire marshal on May 1, 1937, prompting his resignation from the Providence force. That same year the International Association of Fire Chiefs recognized Charlesworth's efforts in fire prevention by appointing him a national director. In 1940, at the age of sixty-eight, he was picked to organize a firefighting force at the Quonset Naval Air Station, and later he served as a local advisor to the navy.

Charlesworth never lost interest in his men and the department that he so proudly served

for four decades. His peers responded by electing him president of the Rhode Island Fire Chiefs Club for twenty years. Charlesworth died on October 12, 1958, at the age of eighty-five. Photograph, 1916, courtesy of Rhode Island Historical Society, Rhi x3 5022.

Providence's crowded waterfront area presented particular problems for the city's firefighters. Lack of waterborne fire equipment had severely hampered firemen in their effort to control the Providence Coal Company fire in 1889. A report issued by the National Board of Fire Underwriters in 1905 pointed to the potential hazard of a major waterfront fire and recommended "that arrangements be made whereby the city may obtain the services of one or more

tugboats, equipped with powerful fire pumps, standpipes and hose, for use at all waterfront fires." The recommendation fell on deaf ears.

On September 17, 1923, a fire at the Morris and Cummings Dredging Company resulted in a loss of more than $200,000. Three weeks later a major conflagration at the Waterfront Coal and Grain Company on Dyer Street caused a $500,000 loss. In November the fire department renewed its effort to secure funding for the

purchase of a fireboat, but the City Council again shelved the request.

Improvements, however, were authorized for the Providence harbormaster's boat. Pictured here are members of Hose No. 15 on Wickenden Street in 1924 demonstrating its possible use as a fireboat. Photograph, 1924, courtesy of the Providence Journal.

tion of Mayor James E. Dunne secured approval for the $200,000 sale in April 1935. In the following year the school department, which had maintained headquarters in the third story of the building, removed to Technical High School and the fire department took up temporary (and very cramped) quarters in the old Richmond Street station.

The Central Station was razed in 1938 to make way for the present Federal Building Annex (the Pastore Building), and the city purchased the former LaSalle Academy property as the location for its new Bureau of Police and Fire Building. That structure, which still serves as departmental headquarters and was home to the "Three Ones," was completed and occupied in 1940. Coinciding with the opening, the *Journal* published a story inventorying the city's old volunteer fire stations. Only one remained—the building of the turbulent Pioneer Engine Company at 296 South Main Street, opposite Hose 2.

As the fire service experienced a succession of headquarters, it also experienced a succession of chiefs. In May 1937 Charlesworth ended a distinguished fourteen-year tenure to accept an appointment as Rhode Island's first fire marshal from Governor Robert E. Quinn. John H. Fischer, Charlesworth's assistant, a forty-year veteran and the oldest firefighter in point of service, was awarded a brief (six-week) tenure as chief, an honor similar to that once accorded Holden Hill. Following Fischer's retirement, his younger Washington Park neighbor Thomas H. Cotter, head of the recently created training school, succeeded to the post and began a productive fifteen-year regime.

To accommodate pensioners such as Charlesworth, Fischer, and approximately 110 surviving rank and filers, the retired members of the department formed an organization in November 1937 "for mutual benefit in sociability and comradeship," claiming (perhaps extravagantly) that it was "the first of its kind known in the United States." Jeremiah McDonald was chosen by the group as its first president.

World War II erupted in September 1939 and exerted a significant impact upon the fire service, especially after the institution of the first peacetime draft in 1940. As firemen were called into the armed service (even Mayor Dennis J. Roberts was activated), an auxiliary corps of fire volunteers was recruited in March 1942 as part of the Civilian Defense Program. The auxiliaries were initially divided into twenty-four squads of eight men and were assigned for evening duty to each of the twenty-four permanent hose companies in the department.

As a precaution against industrial sabotage, such as occurred during World War I, a Greater Providence Water Front Activities Committee was formed in 1942. The creation of this unit, on which Cotter served, failed to prevent the disastrous fire of December 31, 1942, which destroyed the fabricating plant at the new Rheem Shipyard (later Walsh-Kaiser) at Field's Point where the steel plates for the yard's Liberty ships were shaped and cut. Although an FBI investigation revealed no enemy involvement, greater safety measures at this strategic site were necessary. The Rheem fire, which destroyed large quantities of valuable machinery, caused damage estimated at $1.7 million, making the disaster the greatest in dollar amount of any fire in the city's history.

Other notable conflagrations during the war years were the Point Street Grammar School fire of January 1940, which inflicted $500,000 worth of damage and injured the principal, three pupils, and six firemen; a $150,000 blaze in February 1941 that destroyed the four-story brick freight office building at the east end of Union Station, injuring six firemen; and a five-alarmer that leveled historic Infantry Hall on October 4, 1942, less than five hours after the official start of Fire Prevention Week. The latter blaze—which produced a $160,000 loss and injured two firefighters—deprived the city of a once famous civic auditorium. Since its opening in 1880, this assembly hall of the First Light Infantry had served as a forum for three presidents (McKinley, Teddy Roosevelt, and Taft), as a concert hall for a number of famed musicians, and as the scene of such diverse events as prize fights, dog shows, banquets, roller polo matches, and political rallies.

The war years also witnessed several major organizational and structural changes in the department. In 1940 Cotter created a twelve-man diving crew, which he believed to be "the first life-saving organization of its kind to be formed in a fire department in the country." In 1942 an

Fears about the possibility of another catastrophic waterfront fire were realized during the early morning hours of February 25, 1931. State Pier No. 1 had served since 1913 as a major destination for steamship lines carrying immigrants, principally from Portugal and Italy. The twin-towered pier and shed, located east of Allens Avenue and south of Public Street jutted six hundred feet into Providence harbor. The entire structure was supported by thousands of creosote-coated wooden pilings.

The first alarm was pulled at 3:59 A.M. by police sergeant who spotted smoke and flames billowing from the roof of the building. Firemen arrived to find the southern end of the structure enveloped in a wall of fire. Soon the heat began to ignite the building's support columns. Pumpers were placed on the tugs Maurania and Gaspee to assist the harbormaster's boat in trying to prevent the fire from consuming the pier's pilings. A team of firemen from Hose No. 2 joined the fight by manning a raft, and at the height of the blaze all five men were thrown into the water almost directly under the pier. Only quick work by the harbormaster's crew averted tragedy.

By 8:00 A.M. intense heat began cracking the cement walls and flooring of the pier. Shortly after, Chief Charlesworth ordered all firemen from the southern end of the structure. Less than fifteen minutes later the entire southern side sank into the harbor with a thunderous crash.

Twelve hours after it began, one of the city's most disastrous waterfront fires was under control. The loss was estimated at nearly half a million dollars. Photograph, February 25, 1931, courtesy of Rhode Island Historical Society, Rh (L866) 916.

This sixty-five-foot-long aerial ladder truck (left) was added to the Providence Fire Department in October 1926. Its unveiling, according to one news account, caused "an awe stricken citizenry to gape and wonder." The truck became Ladder 1, working out of the Central Fire Station. In 1935, after the station was closed, Ladder 1 was transferred to Hose 7 on Richmond Street while plans were developed for the construction of a new headquarters at LaSalle Square. City work crews had to remove a hydrant and a portion of the sidewalk on the westerly side of Richmond Street to accommodate the mammoth vehicle.

In 1937 a new aerial ladder truck was purchased and old Ladder 1 was sent to the Police and Fire Paint Shop on Bucklin Street for complete refurbishment. She emerged from the paint shop as Ladder 10 (lower left) and was assigned to the Point Street station in South Providence. Photograph, 1935, courtesy of the Providence Journal. Photograph, 1938, courtesy of the Providence Fire Department.

The decades between the wars (1919-1939) have been hailed as the Golden Age of American Sports. However true that designation, these years certainly marked the high point of athletic activity for Providence's safety services. Boxing, wrestling, and baseball contests between the police and fire departments were frequent and exciting. These events attracted large crowds and drew the attention of sports cartoonists like the Journal's Paule Loring. This Loring cartoon of May 1932 indicates the popular interest in an upcoming middleweight boxing bout between

"smoke eater" Frank Moise, an amateur boxing champion, and patrolman Gene Zedick. This slugfest was the feature event of a four-hour police-fire competition in boxing and wrestling that drew nearly five thousand spectators to Rhode Island Auditorium on the evening of May 2, 1932. The cops edged the firefighters 23½ to 18½, with Zedick narrowly outpointing Moise. Journal sports reporter J. D. McGlone saw that match differently: "In case anybody asks me in private what I thought of the Zedick-Moise decision," said McGlone, "I'll probably

tell them—not out loud, you understand—that it looked as though Moise had the first three rounds and should not have been forced to box an extra round. If Moise had been declared the winner, the firemen would have won the tournament, if that is any consolation to them until they make another effort next year." Cartoon by Paule Loring, courtesy of the Providence Journal and Chief Michael Moise.

Unfortunately, too few of these formal photos of early fire station crews have survived. This group picture of Hose 18 on Broad Street was taken in 1935. The leaders of this contingent were Captain Francis H. Morton, Jr. (front row, third from left) and Chief Joseph Dorsey (front row, third from right). One distinguishing feature of this unit was the number of future Providence

firefighters that it produced. Dorsey was the father of Austin and Joseph F. Dorsey, the latter of whom died in 1963 from injuries suffered in a Smith Hill blaze. Frank Moise (back row, extreme left) was the father of William Moise, firefighter and councilman, and Chief Mike Moise; George H. Lowe (back row, second left) was the father of firefighters Arthur and George, Jr.

James L. McNamee, father of this photo's donor, is the tall Irishman in the back row, third from the right. This station is presently the Washington Park Branch of the Providence Public Library. Photograph, 1935, courtesy of Margaret McNamee Sarault.

John H. Fischer's tenure as chief of the Providence Fire Department spanned only thirty-five days, but his career with the fire service encompassed four decades. When Fischer joined the department in 1897 as a hoseman for Hose Company 19 on Plainfield Street, his father was serving as captain of Niagara 5 on Olney Street. Young Fischer was named lieutenant in 1906 and five years later rose to the rank of captain. He was present at the tragic Washington Bowling Alley fire, which killed four of his comrades. When he was appointed batallion chief on March 3, 1923, Fischer took charge of the First Fire District. A year later Chief Charlesworth named him to be his assistant, and Fischer held this position for the next fourteen years. At a fire in 1934 the assistant chief had both hands frozen while directing his men in subzero temperatures.

Fischer became chief on May 3, 1937, and held that post until June 8. He died on December 27, 1942, five years after his retirement. Photograph, 1916, courtesy of Rhode Island Historical Society, Rhi x3 5021.

On April 23, 1935, the Providence City Council voted to sell the Central Fire Station to the federal government for $200,000. Its site was to become the location of a new federal post office. Two months after the sale the Board of Public Safety was superseded by the Bureau of Police and Fire. City planners soon began the process of seeking a Downtown site for a new combined police and fire station. In 1938 the former LaSalle Academy property on Fountain Street was purchased, and two years later the Bureau of Police and Fire Building opened. Engine

Company No. 1 and Ladder Company No. 1, temporarily housed at Hose 7 on Richmond Street, were moved to their new home on LaSalle Square along with the department's administrative offices and the Bureau of Fire Prevention.

Ladder 1, shown here on the right, was purchased in 1937 and was the first piece of new equipment added to the department in ten years. Photograph, circa 1940, courtesy of the Providence Fire Department.

Thomas H. Cotter's career with the Providence Fire Department spanned nearly a half century. Born in Providence in 1877, Cotter first worked at the E. L. Freeman Company and the Gorham Manufacturing Company. He applied for appointment to the fire department in 1902 and was given a job as a substitute two years later. In September 1904 the twenty-seven-year-old Cotter was assigned as a ladderman to Ladder Company No. 5 on the corner of Public and Burnside streets in South Providence. Eleven years later he was transferred to the Point Street station.

By 1918 Cotter had risen to the rank of captain. Five years later he was named battalion chief for the Second Fire District, which comprised much of South Providence and Washing-

ton Park. In 1929 the Board of Public Safety authorized him to attend New York Fire College in preparation for his proposed duties as head instructor of the newly created fire training school on Bucklin Street. Named deputy chief in 1936, Cotter succeeded Chief John H. Fischer a year later, becoming the first Irish-American to head the department.

In 1941 Cotter launched a campaign to establish a rescue unit within the department, and within a year the City Council authorized him to purchase the department's first rescue vehicle. Under Cotter's direction the fire service initiated the largest station-replacement program in its history, beginning work on nine new stations. A reorganization of the force reduced the number of fire companies from thirty-seven to thirty. In

June 1951 illness forced Cotter to apply for a leave of absence. He retired in October, turning over the reins of the department to his capable deputy chief, Lewis Marshall. Cotter died in 1961 at the age of eighty-six.

Shown here with Cotter (left) is Marshall (center) and Thomas H. Roberts (right). Roberts, brother of Mayor Dennis J. Roberts, was one of the first members of the Bureau of Police and Fire when that body was created in the aftermath of the "Bloodless Revolution" of 1935; later he served as chief justice of the Rhode Island Supreme Court (1966-1976). Photograph, courtesy of the Providence Fire Department.

In an effort to reduce the incidence of water-related fatalities, the Providence Fire Department in 1940 organized a team of twelve firemen and provided them with a summer-long program of instruction in both saltwater and freshwater diving. Pictured are the members of the Providence "Diving Crew" receiving sweaters from Chief Thomas Cotter. Shown left to right are (first row) Alfred B. Butler, Francis H. Higney, Charles W. Oatley, John Feeney, John F. Kirkconnen, and Charles F. Potter; (back row) James H. Coleman, Arthur Brodeur, George H. Nowell, William J. Quirk, Edmond H. Boyes, and Frederick P. Cooney. This team was touted by Cotter as the "first life saving organization of its kind to be formed in a fire department in the country." Photograph, October 4, 1940, courtesy of the Providence Journal.

official rescue squad (first proposed in 1929) was commissioned and given a "catastrophe wagon," described by squad members Charles F. Potter and John Feeney as a "mobile hospital-tool shed-carpenter shop-power plant." Five years later the department added two "rescue boats" to cope with water accidents. As usual, Providence continued to maintain its leadership in the field of fire prevention, winning first-place awards from the U.S. Chamber of Commerce in 1940, 1942, and 1944.

The most controversial issue involving the department during this era was the union movement. In 1935 Congress passed the National Labor Relations Act upon the urging of Senator Robert Wagner. This measure guaranteed labor the right to organize and bargain collectively. In the years following passage of the Wagner Act, unions flourished. By the mid-1940s this wave of unionization spread to public-service employees like the Providence firefighters. In July 1944 three hundred men petitioned Chairman Edward L. Casey of the Bureau of Police and Fire to amend Rule 49 of the department regulations to permit formation of a firemen's union in Providence. The petition expressed concern "that with the return of our troops and their demobilization, problems of finding employment for them will confront each city. Shorter hours will be a factor in solving that problem, also the adjustment of salaries to meet living costs, the adjustment of pensions, and the security of our positions, will all be problems that can best be met by having an efficient organization with competent officers, who have given some thought to the study and solution of such problems."

Mayor Roberts and city officials at first did not agree. All four of the proposed union officers—Raymond O. Baggesen, Joseph J. Ellis, William J. Hughes, and Joseph J.

Trudeau—were transferred or exiled to Hose 20 on Manton Avenue after a disciplinary hearing "for actions which were prejudicial to department discipline." The Democrats, it seemed, favored unionization only when it did not interfere with political patronage.

The International Association of Firefighters sent an official from its Washington headquarters to help fight the city ban, and in early 1945 unionism prevailed. Rule 49 was amended to allow privates to participate in the formation of a labor organization. In October 1945, however, when officers applied for an exemption from Rule 49, Chief Cotter successfully and firmly blocked the effort, alleging that an officers' union "would disrupt departmental discipline."

By the time Cotter exercised his veto, he had been given more power over personnel practices than any other chief before him. In March 1945 Mayor Roberts and the Bureau of Police and Fire had announced a sweeping reorganization of the department that would formally establish a Division of Training, or "Fire College," provide instruction for all recruits, and vest control of appointments and promotions "fully and exclusively" in the chief. This effort to eliminate the political interference in personnel matters that had plagued the department for nearly two decades was designed, said the bureau, "to place appointment and promotion under such a system of controls that it becomes impossible for any person, in or out of the department, to take any positive action to gain appointment or promotion for an individual, and to leave the sole power to take negative action...within the exclusive authority of the Chief of the department to be exercised only at such times as the best interests of the department require." The *Providence Journal* editorially hailed the plan as one designed to create "a department

Union Station has been a landmark for more than eight decades. Originally built as a five-building complex, that number was reduced by one on the evening of February 18, 1941. Just after 9:30 P.M. the easternmost building, housing the Railway Express Agency and office space of the New Haven Railroad, caught fire. Within ten minutes the entire four-story structure was enveloped in flames that could be seen as far away as Woonsocket. Six firemen were injured in this three-alarmer, and several pieces of fire equipment, including Ladder 1, were scorched by the intense heat. The building was declared a total loss and razed a short time later. Plans are now being developed to reconstruct this edifice as part of the Capital Center Project in order to restore the original symmetry to Union Station. Photograph, February 19, 1941, courtesy of Rhode Island Historical Society, Rhi x3 4932.

built and run on the principle of merit alone."

Relentless union pressure brought about the final significant personnel change of the Cotter era in January 1947. This reform reduced the duty week from eighty-four hours to sixty-eight. The requirements that firemen gain permission to leave town on off-days and be on call for "three-baggers" were eliminated. Such improved working conditions prompted a departmentwide "Farewell Party to the Long Work Week." The new schedule, which also reduced the longest duty period from twenty-four to fourteen hours and gave the men forty-eight hours off every ten days, necessitated the addition of eighteen new recruits to the force.

In these last years of Cotter's productive tenure another station-building boom occurred. This development, spearheaded by Mayor Roberts, began in 1945 with the passage of a $1,125,000 bond issue by the council and the state legislature for the "modernization of the housing and other facilities" of the department. By September 1948 bids were let for the construction of the first four (of nine) new fire stations to be financed by the bond issue. In March 1950 Hose 18 was installed at the new Allens Avenue station near Ernest Street and Hose Company 8 and Ladder Company 2 were transferred from their eighty-five-year-old Harrison Street building to the new facility at Messer Street and Union Avenue. By 1952 the remain-

ing seven stations were in service: Broad Street, Hartford Avenue, Admiral Street, Branch Avenue, Atwells Avenue, Brook Street, and North Main at Meeting Street. These nine modern stations were functional in design, but much less picturesque than the bell-towered brick or brownstone structures they replaced. In two instances stations were built at the expense of valuable historic structures that had occupied their sites. The North Main Street facility caused the removal of a 1725 Quaker meetinghouse, and the Brook Street station was built by razing a house constructed in 1830 by Samuel Ames, famous chief justice of the Rhode Island Supreme Court. Progress had its price.

In 1951, with the building-modernization program in full swing, the department experienced a significant administrative shake-up. Walter Reynolds began a fourteen-year tenure as mayor, succeeding Dennis Roberts, who advanced to the governorship; the Bureau of Police and Fire was abolished in favor of a mayorally appointed office of public safety commissioner, with John B. Dunn its first incumbent; and an ailing Chief Cotter took sick leave in June, entrusting a department consisting of 471 men, thirty-two pumpers, and eighteen ladder trucks to the care of Lewis A. Marshall. With new facilities and new leadership, the fire force embarked upon a new and modern era of service to the city.

Rescue squads were first organized in Chicago in 1913. Initially, rescue apparatus was used to carry extra firefighters to the scene of blazes, but soon equipment was added to allow firemen to render first aid and make forcible entry into burning buildings. In 1915 New York established a ten-man squad and equipped its truck with oxygen tanks and air masks.

In 1938 a detail of Providence firefighters led by Charles F. Potter was sent to New York for a three-week training session with the rescue units there. Almost four years later, in January 1942, at the urging of Chief Cotter, Rescue 1 became a permanent unit of the Providence Fire Department. Private Frederick L. Badger recalled that the first run of the squad was to a meat-packing plant on Canal Street to stop an ammonia leak.

The four-man squad, with its 1939 Chevrolet truck, answered 254 calls in its first year of operation. Increased demand led to the formation of a second rescue unit in April 1952. This photograph shows the original rescue truck with the new Rescue 2 vehicle that the department purchased in May 1954. In 1957 a third rescue unit was formed, bringing the total strength of the fire service component to eighteen men. By 1970 the rescue squads were answering more than ten thousand calls annually. Photograph, 1954, courtesy of the Providence Journal.

On the night of October 4, 1942, Infantry Hall, a Providence landmark for more than sixty years, succumbed to flames. Located on South Main Street, this four-story brick building contained a two-thousand-seat auditorium that had served as a forum for three presidents and as a theater for such attractions as the American Band, Irish tenor John McCormack, violinists Jascha Heifetz and Fritz Kreisler, and Providence's own Sissieretta Jones ("Black Patti"), a noted entertainer.

The fire broke out just before 5:00 A.M. Within a short time the flames extended along the entire two-hundred-foot width of the building. In all, five alarms were sounded, the third bringing in members of the day platoon. Twenty-one hose companies, ten ladder trucks, and the department's water tower (visible at the extreme left of the photo) were committed to extinguish the blaze. Fire damage was limited to the third and fourth floors, but the venerable old building never recovered from its bout with fire and soon felt the wrecker's ball. Photograph, October 4, 1942, courtesy of the Providence Journal.

The Infantry Hall fire was just a prelude to the city's most costly single fire loss in its history. Just after two o'clock on the afternoon of December 31, 1942, workers at the Rheem Shipyard's fabrication plant off Allens Avenue, near the city line, noticed smoke billowing from the pump house at the east wall of the huge 640-foot-long building. In the plant at the time were several hundred workers cutting steel plates used in the construction of Liberty ships being produced by the yard. The 156,000-square-foot multistory structure had been sheathed in plywood because of the unavailability of sheet metal.

Flames vaulted the east wall, and within minutes the fire was out of control. Workers in the plant literally ran for their lives. Compounding the problem was the fact that sprinkler equipment that had arrived at the plant five months earlier had not been installed. By three o'clock more than thirty pieces of fire equipment from Providence and nearby Cranston were battling the enormous blaze. Thick black smoke rising from the plant could be seen as far as Fall River. In Providence members of twelve auxiliary fire companies were summoned to stand by.

Almost within an hour all that remained of the building was its steel framework. The loss was later estimated at a staggering $1.7 million, including valuable production equipment. Miraculously, no one was seriously injured. Photograph, December 31, 1942, courtesy of the Providence Journal.

New York's terrible Triangle Shirtwaist fire in 1911 jolted public awareness into the need for a national effort aimed at eliminating dangerous conditions that often resulted in tragedy. In 1916 a national effort was launched by the United States Chamber of Commerce to set aside a day in October as Fire (or Accident) Prevention Day. In Rhode Island both the governor and Mayor Gainer of Providence issued proclamations urging the citizenry to clean and remove "all rubbish and inflammable material from their premises."

It was not until 1922—on the fifty-first anniversary of the Great Chicago Fire—that fire prevention gained nationwide prominence. Warren Harding's presidential proclamation established October 2-9 as National Fire Preven-

tion Week. Providence marked the occasion with a number of activities spearheaded by the local Chamber of Commerce.

Heightened interest led to a call for the establishment of a fire prevention unit within the fire department. Again, it was the chamber's Providence Safety Council which led the charge. On December 23, 1925, the Providence City Council authorized the Board of Fire Commissioners to organize a Bureau of Fire Prevention. The ordinance empowered the bureau to "investigate the cause, origin and circumstances of every fire occurring in the city"; to supervise the transportation of hazardous substances; and to inspect at least four times a year "all especially hazardous manufacturing processes, storage or installations of acetylene or other gases, chemical oils,

explosives and inflammable materials [and] all interior fire alarm and automatic sprinkler systems."

The Bureau of Fire Prevention, Fire Prevention Week, and spectacular public displays soon became permanent fixtures in the effort to reduce the incidence of fire in the capital city. One photograph depicts members of a special twenty-man fire department squad performing at the finale of Fire Prevention Week atop eighty-five-foot-high aerial ladders in Roger Williams Park on October 9, 1942; the other shows firefighters scaling the Chamber of Commerce building on Westminster Street in the late 1940s. Photograph, October 9, 1942, courtesy of the Providence Journal. Photograph, circa 1947, courtesy of the Providence Fire Department.

On the evening of November 13, 1947, the Rhode Island Recreation Center was packed with more than four hundred bowlers, many of them from the Temple Beth-El bowling league. Next to the forty-two-alley center on North Main Street near the Pawtucket line, the staff at Sullivan's Steak House was entertaining the evening's dinner guests. In the kitchen the chef on duty had begun to broil a half pound of bacon just before 10:00 P.M. when it suddenly ignited. The oven exploded, setting the ceiling ablaze. The fire spread quickly to the Muriel Room, a section of the bowling alley. Within minutes a blast of heat and flame swept the main alleys. Orderly retreat soon turned to desperate panic. Bowlers jammed the exits, while others smashed plate-glass windows and made their escape. When firemen arrived, flames were bursting from the center's doors and windows.

At 10:22 P.M. a third alarm was sounded, triggering air raid sirens that called all off-duty firemen to their stations. At 11:07 the sirens were put to work again sounding the fifth alarm. Several minutes later the roof of the Recreation Center collapsed. Thirty-nine injured patrons were rushed to area hospitals, but amazingly no one was killed. The following evening Rabbi William Braude led a special service of his congregation to thank God for the "safe delivery" of his temple members. Photograph, November 14, 1947, courtesy of the Providence Journal.

A drill school had been established by the department in 1929 with the completion of the seven-story drill tower off Bucklin Street (below left). In the fall of 1945 a formal training program for new firemen began. Located first on Central Street and later on Reservoir Avenue, the Division of Training divided the six-month program into thirteen weeks of basic and unit training followed by another thirteen-week period of advanced training in the station house. Promotional classes were also established for privates, lieutenants, and in 1968 for captains.

In this photo (below right) Chief of Training Lewis Marshall instructs trainees in the techniques of fighting fires in the Downtown. The model, constructed by the carpenter shop, was an exact replica of the Providence city block bounded by Westminster, Mathewson, Weybosset and Union streets. Photograph of drill tower, courtesy of the Providence Fire Department. Photograph, circa 1947, courtesy of Chief Michael Moise.

Between 1948 and 1952 the department experienced its last major station-building boom. All stations antedating 1903 were closed to active use, and nine modern stations were erected. Four of the stations—on Broad Street, Allens Avenue, Messer Street, and Hartford Avenue—were designed by Lucio E. Carlone; the others—on Admiral Street, Atwells Avenue, Branch Avenue, Brook Street, and North Main Street—were the work of Jackson, Robertson, and Adams. The prime movers behind this modernization effort were Mayor Dennis J. Roberts and Chief Thomas H. Cotter. Lewis Marshall brought the project to a successful conclusion in 1952.

Shown here is the station of Engines 6 and 7 and Ladder No. 4 at North Main and Meeting streets shortly after completion at the start of a spring cleanup campaign. The structure replaced a 1725 Quaker meetinghouse that formerly occupied the site. Photograph, circa 1955, courtesy of the Providence Fire Department.

Lewis Marshall exerted a profound and lasting impact on the quality of Providence's fire service. Few will deny that he ranks among the city's most important chiefs. Marshall began his career in 1924 as a hose-wagon driver for Engine 7 on Richmond Street. Promotions quickly followed. By 1945 the Providence fireman had attained the rank of battalion chief.

Earlier he had helped establish the Bucklin Street training school and added classes in hydraulics and fire engineering. Marshall's administrative and training talents soon drew national attention.

Appointed chief in 1951, Marshall continued the physical reorganization of the fire depart-

ment. Of the thirty fire stations in use when he took office, twenty-three were soon abandoned, having been replaced by nine new strategically located facilities. Marshall also embarked upon an ambitious apparatus-replacement program and changed the department's two-platoon sixty-eight-hour work week system to a three-platoon system with a fifty-six-hour week.

Throughout his tenure Marshall's department won national awards in fire prevention. A strong believer in community involvement in this area, Marshall attributed Providence's success to public education by the department and the fact that "a good department comes directly from an intelligent and cooperative

public." Marshall himself was honored in 1962 by being elected president of the International Association of Fire Chiefs—the second Providence chief to win that distinction.

A quiet man, Chief Marshall silently endured the effects of a painful back injury suffered when he fell from a ladder while fighting a fire in a box factory on Chestnut Street in the 1930s. This injury, aggravated by several others, forced his retirement in January 1967. Marshall died ten years later on August 7, 1977. Photograph by Earl Goodison, courtesy of the Providence Fire Department.

chapter FOUR

The Modern Era, 1951-1984

On October 18, 1951, forty-eight-year-old Lewis A. Marshall, acting head since June, was sworn in as chief of the fire department by Public Safety Commissioner John B. Dunn. Few men came to the post so well equipped. The serious and businesslike Marshall had served in all branches of the fire service since his appointment as a recruit in 1924. First a hoseman, then captain of Ladder 10, then drillmaster of the department, he became deputy chief in 1949. In that capacity he served as consultant to the architects who drew the plans for the city's new fire stations. Despite his administrative duties, he had found time to attend a number of special schools for fire department administration and firefighting techniques. From October 1951 to January 1967, Providence would have its own Marshall Plan.

The Marshall era was truly modern in many respects. The last group of new stations were opened under his direction while he simultaneously supervised the phasing out of twenty-three older houses, putting most on the auction block. In July 1955 he led the successful move for a three-platoon system on a nine-day cycle that reduced the work week from sixty-eight to fifty-six hours. Despite his image as a disciplinarian, that reform endeared him to the rank and file.

With the approval and support of Mayor Reynolds, Commissioner Dunn, and Dunn's successor Francis A. Lennon, Marshall instituted a major program of apparatus replacement and established a large reserve fleet of pumpers and ladder trucks. Between 1952 and 1954 the department acquired nine new pumpers for its fleet, and when Hurricane Carol inundated Downtown Providence on August 31, 1954, placing "a severe strain upon the personnel and equipment of the fire department," Marshall's policy of retaining apparatus that had been replaced proved its value as the reserve equipment was used to great advantage in pumping out the business sections of the city. "With 15 reserve pumpers and our regular equipment, plus 20 Homelite pumps, we had the water situation cleaned up within a very short time," boasted the department's annual report.

Another area of growth was the rescue fleet. To take the increasing load off Captain Fred Badger's harried crew at La Salle Square, Rescue 2 was created in 1952 and placed under the command of Lieutenant Arthur Brodeur. Stationed at Messer Street, the six-member unit was charged with servicing all calls from Olneyville, Manton, Mount Pleasant, Elmwood, and Washington Park. Marshall justified its creation on the grounds that rescue runs had increased from 254 in 1942—the first year of operation—to 700 annually by 1951. At first the new squad had to be content with Rescue 1's castoff 1939 "catastrophe wagon," but in May 1954 Marshall presented the unit with a new truck. In 1957 Rescue 3 was formed as annual runs increased to more than 4,700. Led by Lieutenant Joseph Healy, the newest squad called the Branch Avenue house of Engine No. 17 and Ladder No. 9 its home.

Another major piece of modern apparatus obtained in the early 1950s went to Ladder Company No. 9. The features of this new $37,500 American-La France aerial ladder were extolled by the *Providence Journal* in an October 1952 pictorial news essay. The device could be raised 85 feet, a height of six 14-foot stories, allowing a firefighter to bring a heavy stream of water directly upon a fire raging 80 feet above street level. The pride of Ladder 9 was built in three sections (called flies) which telescoped into one another. A hydraulic lift, powered by the truck motor, raised the ladder into firefighting position and lowered it. One man could easily operate the device, leaving the other members of the company free to fight the blaze. Such modern design was a far cry from the Skinner contraption used during the Steere regime. By 1963 a 100-foot Maxim steel extension ladder was also in use.

The ultimate in modernization came in 1961 when the department adopted a new siren. Through a flip of a dashboard button the driver could cause either a continuous wailing or a shrill, high-pitched "yelp," which Marshall justified on the grounds that "it was more penetrating than the conventional signals" then in use. He might have added that it was worlds apart from the stationary tower bells of the mid-nineteenth century.

The most striking aspect of departmental performance in the period 1952 through 1971, and a source of much civic pride, was the record of awards won by Providence in National Fire Prevention Week contests. During that twenty-year span the city ranked first in its population class eleven times (it was never lower than fourth), including a six-year streak from 1959 through 1964. In 1962, 1968, and 1971 it won the Grand Award as the nation's fire prevention leader. By 1966 the American Insurance Association's periodic ratings placed Providence among sixteen U.S. cities with a 2A rating. No other community ranked higher.

Much of the credit for this impressive performance must go to the department's twelve-man Bureau of Fire Prevention. Its director, Battalion Chief Leo E. Gorman, initiated these two decades of achievement in 1952, when

This photograph depicts the dramatic effects of Chief Marshall's modernization program. The triple combination pumpers of 1920s vintage (right) were phased out and placed in reserve. The two new 750-gallon-per-minute cab-ahead-of-engine pumpers (left) were part of a fleet of eight new pieces of apparatus purchased in 1952 and 1953. One of the two American-La France trucks pictured here was assigned to the department's Brook Street station (background), which had opened in 1950. Photograph, June 19, 1952, courtesy of the Providence Journal.

Providence won first prize among comparable cities for "the most notable fire loss record of 1952," the "best firefighting facilities," and "the best community educational programs." In succeeding years investigations, demonstrations, lectures, radio and TV broadcasts, fire drills, posters, pamphlets, and good old-fashioned boosterism kept the Providence Fire Prevention Week program always at or near the top. Gorman was especially proud of the fire inspection program, claiming in his 1958 report that Providence was "the first metropolitan city in the United States to conduct dwelling inspections by uniformed firemen." After Gorman's death in 1960, his successor, longtime bureau member John E. Butler, brought the city to even greater heights when it earned the Grand Award in 1962.

Another important development of the Marshall years (though accomplished despite the opposition of the Reynolds administration) was the strengthening and solidification of the union movement. The crucial test came in May 1962, a year after the state Fire Fighters' Arbitration Act gave the firefighters' union "all of the rights of labor other than the right to strike," but including the right to a closed shop. The bargaining representation election of May 14 was staged by Captain John F. McDermott, who claimed that it was "the first of its kind among unionized firemen in the nation." When the balloting at the Division of Training was done, 431 of 446 eligible union members had voted, and all of them authorized Local 799 of the Fire Fighters' Union, AFL-CIO, to act for them as their collective bargaining agent with the city.

Almost immediately negotiations for the firefighters' first union contract began, but the talks dragged through the summer and into early fall. When union and city officials reached an impasse in July, the contest went to arbitration, with Salvatore DiSano acting as the union representative. Finally, on October 30, 227 firemen (captains, lieutenants, and privates), meeting in the basement of Elks Auditorium on Washington Street, gave approval by a 6 to 1 margin to an agreement that included a four

dollar-per-week pay raise and a small allowance for uniforms. From that year onward, Local 799 has been a potent force in shaping the wages, hours, and working conditions of the Providence Fire Department. By November 1966, for example, the union succeeded in again reducing the work week for city firemen from fifty-six to forty-eight hours.

During the final years of Marshall's tenure, several important administrative changes occurred. In 1963 the rules and regulations of the department were completely rewritten for the first time since 1927; an air supply station was established in 1964 to repair and maintain all self-contained breathing apparatus and charge all air tanks and dry-powder extinguishers; Edmond F. Marnane (later a leader of the retirees) became the new head of the Division of Training; James T. Killilea assumed the direction of the Fire Prevention Bureau in 1965 upon the retirement of Captain Butler; and the Fire Alarm Division was placed under the able supervision of Alfred J. Mello, who had succeeded longtime superintendent Henry Van Westerndorf. The youthful Mello immediately set to work modernizing the Kinsley Avenue communications center until he could claim without exaggeration, as he did in his 1969 report, that Providence now had "one of the most modern fire alarm centers in the country." (In 1972 communications was constituted a separate department of city government, and Mello is still its director.) In the midst of these changes in the fire service, Joseph A. Doorley, Jr., son of a Providence fireman, succeeded Walter Reynolds as mayor in January 1965 and promptly installed political associate Harry Goldstein as Public Safety Commissioner.

Marshall's last years on the force brought him the honors that his lifetime of fire service had merited. In 1963 he presided as president of the International Association of Fire Chiefs, following in the footsteps of his predecessor George Steere. By 1963, however, the association had a membership of eight thousand fire officials

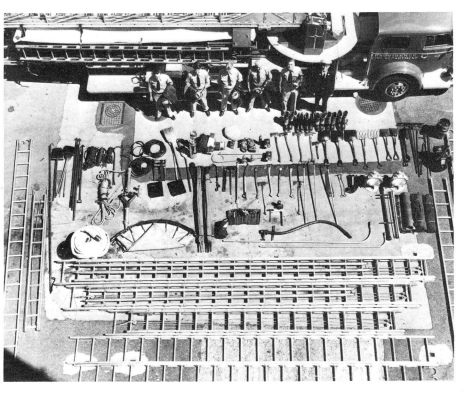

Ladder 9, operating out of its station on the corner of North Main Street and Branch Avenue, represented the department's frontline defense against fire. Purchased in 1952 as part of the modernization program, the American-La France Foamite steel aerial truck boasted a three-flie extension ladder that could be raised eighty-five feet. A hydraulic lift, powered by the truck's motor, raised the ladder into position. A ladder pipe with a nozzle could be attached to the top rung of the extension to allow firemen to play a straight stream of water on high-rise fires.

The photo depicts members of Ladder 9 with a full load of equipment spread out before them. Items include seven extension ladders, a life net, a gooseneck pipe for shooting water into cellars, fifty-foot sections of three-inch hose, salvage tarpaulins, and two gas masks. Chief Marshall is shown at the extreme right. Photograph, 1952, courtesy of the Providence Journal.

It wasn't long before the department's reserve equipment was put to good use. On August 31, 1954, Providence was surprised by an unwelcome visitor named Carol. The hurricane slammed the city with gusts of over one hundred miles per hour. As in 1938, the storm hit at high tide and was accompanied by tidal waves. The Downtown was flooded with waters reaching thirteen feet above mean high water, just nine-tenths of a foot below the 1938 mark.

Fire department personnel were mobilized to assist in the cleanup effort. All fifteen reserve pumpers and twenty Homelite pumps were utilized to help local businesses like the Outlet Company empty flooded basements. Photograph, September 1, 1954, courtesy of the Providence Journal.

During the 1950s the department's Bureau of Fire Prevention, under the direction of Chief Leo Gorman, conducted an intensive effort designed at reducing the incidence of fire in the city. Parades, school contests, fire prevention demonstrations, and shows on TV and radio all helped Providence gain a national reputation in cutting fire losses.

Children were a prime target of the department's educational effort. Pictured (preceeding page top left) on WJAR-TV is Ted Knight (later a regular on national television's Mary Tyler Moore Show) with children from the Sackett Street School displaying the message "Clean up, don't burn up." WPRI's well-known "Salty" Brine (preceeding page top right) played his part by inviting Chief Marshall and members of the Wanskuck Boy's Club to "Salty's Shack" to discuss fire prevention. For the grown-ups, the traveling "Hadafire Family" (bottom left) entertained shoppers along Westminster Street with a prevention skit during their visit to Providence.

Other examples of the public relations effort were the 1953 cleanup campaign (top), enthusiastically led by officers of the department, and the 1956 Fire Prevention Week parade (above) through Exchange Place. Photographs, courtesy of the Providence Fire Department.

drawn from all fifty states, from Canada, and from twenty-eight other foreign nations. In addition, Marshall was the only chief on the board of directors of the prestigious National Fire Protection Association, a nonprofit organization devoted to fire prevention and the study of fire engineering. Upon Chief Marshall's retirement in January 1967, a local news story hailed him as "The Father of the Modernized Department."

In the spring of 1966 the General Assembly enacted the most significant piece of firefighting legislation in the state's history. This measure—a comprehensive, two-hundred-page fire safety code—was far more extensive than any preceding state enactment on the subject. Its "underlying purposes," according to the legislature, were "to simplify, clarify and modernize the law governing fires and fire prevention" and "to specify reasonable minimum requirements for fire safety in new and existing buildings" except for private dwellings of three families or less. Cities and towns could still enact ordinances relating to fire safety provided they imposed requirements equal to, or more stringent than, the state code, which is presently designated as Title 23, Chapter 28.1, of the general laws.

Among its many provisions, the code created a Division of Fire Safety under the direction of the state fire marshal and a Fire Safety Code Board of Appeal and Review. It established standards for inspection; set requirements for fire escapes, exit signs, emergency lighting, and fire alarms; and prescribed safety regulations for places of assembly, theaters, health care facilities, schools, hotels, motels, apartment houses, boarding homes, rooming houses, day care centers, industrial and commercial buildings, tents, and grandstands. Finally, the code placed limitations on the use of such combustible materials as fireworks, explosives, liquified petroleum, fuel oil, and other flammable liquids.

Since 1966 such modern items as model rocket engines, liquified natural gas, portable fire extinguishers,

and fire detection systems have been added to the code as well as some more traditional fire concerns such as horse stables. In addition, a new state building code was adopted in 1973.

During the twenty-year reign that Providence enjoyed as fire prevention leader, some spectacular fires marred the record. The first of these occurred on September 22, 1953, when the Tillinghast Plumbing Supply Company on Dorrance Street and four adjacent buildings burned at a cost of $300,000. In January and February 1956 the department reported a rash of fires, "the work of a pyromaniac," causing losses totaling over a half million dollars. On February 5 three multiple-alarm fires (two of them set) produced a state of emergency that prompted Marshall to recall all personnel to duty.

The school department was dealt two devastating blows in 1958. On May 6 Classical High was torched by an arsonist, incurring $150,000 in damages, and on July 26 another "suspicious" fire heavily damaged the school administration building on Pond and Summer streets, destroying many academic records. According to contemporary reports, the latter blaze brought to the site "the largest concentration of fire apparatus to a single fire in this city in thirty-five years." Twenty pumpers and six ladder trucks, two rescue units and a salvage wagon rushed to the scene, where 140 firemen halted the spread of the $600,000 fire.

In the 1960s, big-dollar disasters were suffered by the First Unitarian Church on Benefit Street (August 23, 1966; $475,000); the Shipyard Drive-In Theater (September 4, 1966; $200,000); Federal Dairy on Greenwich Street (January 25, 1967; $200,000); the Eddy and Fisher Wholesale Liquor Company on Charles Street (June 27, 1967; $452,000); the Mount Pleasant Hardware Store on Chalkstone Avenue (June 29, 1967; $238,000); St. Michael's Parish Hall on Prairie Avenue (December 31, 1968; $270,000); and the Jefferson Apartments on Broad Street,

Another National Fire Safety Award is presented here to Chief Marshall and Mayor Walter Reynolds on May 9, 1956. This was just one of a number of awards won by the city in recognition of its excellent record in the area of fire prevention. During Marshall's tenure as chief, the National Fire Protection Association awarded the city top national honors in its population class eight times, and in 1962 Providence was rated first in the country, winning the Grand Award. The city also captured the association's highest honor in 1968 and won it again three years later.

An individual who contributed significantly to the city's rise to national prominence in the fire prevention field was South Providence's Leo Gorman (standing at left between Marshall and Reynolds), head of the Bureau of Fire Prevention in the 1950s. Chief Gorman fathered a son equally adept at organizational and promotional efforts—James "Lou" Gorman, currently general manager of the Boston Red Sox. Photograph, May 9, 1956, courtesy of the Providence Fire Department.

A familiar ritual for fire department members is the competitive entrance examination. Pictured here are some of the 133 applicants who were tested at Central High School on March 5, 1959. Of this record number of applicants, 18 were selected for the twenty-six-week training program. Photograph, March 5, 1959, courtesy of the Providence Journal.

Throughout its long history the department has engaged in numerous civic activities, ranging from fund-raisers for charity to Christmas caroling. In this 1961 scene, firefighters serenade Christmas shoppers near City Hall. Third from the left is Chief Marshall; at the extreme left is young Mike Moise, who is not merely a singer but also an accomplished ragtime pianist. Photograph by Earl H. Goodison, courtesy of the Providence Fire Department.

103

Providence Journal *photographer Edward C. Hanson took this bird's-eye view when the department's new one-hundred-foot aerial truck was delievered in January 1963. The truck was built by the Maxim Motor Company of Middleboro, Massachusetts, at a cost of $45,200. The new vehicle was assigned to Ladder 10 on Point Street, replacing the old wooden Ladder 10 that the department had acquired in 1937. Photograph, 1963, courtesy of the Providence Fire Department.*

The widespread use of cars and the development of interstate highway systems in the 1950s and 1960s resulted in dramatic increases in consumer demand for gasoline and diesel fuels. The transportation of these highly flammable fuels along the state's highways was viewed with increased concern by local fire officials. During the early morning hours of April 18, 1963, their fears were realized. At about 5:15 A.M. an empty oil tanker driven by John B. Preston jackknifed while crossing the rain-soaked Washington Bridge. Preston's truck slammed into a Mobil Oil tanker fully loaded with seventy-one hundred gallons of

gasoline. The force of the crash threw Preston fifteen feet from his cab and tipped the Mobil tanker on its side. Within seconds the Providence side of the bridge was an inferno.

The fire raged out of control for nearly an hour and a half. The intense heat melted the metal skin of the Mobil tanker and the bases of two aluminum light standards. The three-alarm blaze brought eighteen pieces of equipment from Providence and East Providence, including that neighboring city's new fifteen-ton foam pumper, shown here in the thick of the action. The fire closed the bridge to all traffic for five hours, causing one of the worst traffic jams

in the city's history.

Ironically, Chief Marshall had met with Providence Public Safety Commissioner Franci A. Lennon just two months earlier to discuss th danger posed by chemical fires. Two years after the tanker fire the department outfitted a new Seagrave pumper with foam apparatus and placed it in service at old Engine 18 on Allens Avenue near the city's oil storage district. In 1972 Foam Tender No. 1 was added to the department fleet and also assigned to the Allens Avenue station. *Photograph, April 18, 1963, courtesy of the Providence Fire Department.*

February 6, 1969; $203,000).

Much more tragic, however, was the lodging-house fire of February 1, 1962, at 11 Wilson Street, which killed six tenants—the highest death toll to that point in the city's history. March 1963 saw the "needless" loss of Lieutenant Joseph F. Dorsey from injuries incurred while battling a fire in a West Park Street house scheduled for demolition because it lay in the path of proposed Route 95. And in June 1967 Private Earl T. King perished in the inferno that had been the store of Mount Pleasant Hardware. In the 1960s a total of eight firefighters died in the line of duty, a toll higher than that of any decade in the long history of the fire service.

In January 1967 sixty-one-year-old James T. Killilea, of the city's Elmwood section, became the new chief of the department. John T. McLaughlin assumed Killilea's coveted post as head of the Fire Prevention Bureau, and he went to work preparing a two-hundred-page book detailing the city's prevention effort. Included were pictures, press releases, and the letters of verification that earned Providence its third Grand Award in 1968. As part of the fire prevention effort, Mayor Doorley began "Operation Demolition" to rid the city of "hazardous and dilapidated buildings, mostly dwellings, that could never be rehabilitated."

Killilea, who had been both a boxing and a wrestling champion of the department when such competitions were held in the 1930s, made improvements in equipment and apparatus at a pace comparable to that maintained by Chief Marshall. In 1967 newly instituted services included a foam tender and a water pollution unit assigned to Engine 18 on Allens Avenue to combat oil spills. In 1968 the city contracted for two pumpers, a rescue vehicle, and a one-hundred-foot aerial ladder truck, and it placed an aerial lift bucket in service. A year later, using a then novel lease-purchase procedure, the department bought six new Mack trucks at a cost of $297,000.

In July 1970 Chief Killilea ended his brief but productive stint as department head. His successor, South Side battalion leader John F. McDermott, Jr., was the "boy wonder" of the department. At forty-four, McDermott assumed the position of chief at a younger age than any previous chief in this century. His rise to prominence was based on a combination of ability and union clout. McDermott had served four successful and innovative years as president of Local 799.

McDermott took office during one of the "long hot summers" of racial strife that had afflicted the cities of America since the deadly rioting in the Watts areas of Los Angeles in August 1965. Providence—especially the black areas in the South Side and the Camp Street district of Mount Hope—was affected by this recurrent wave of violent protest, but the city fared much better than Los Angeles, Newark, or Detroit. The moderation of the city's black leaders (many of whom had Rhode Island roots predating the arrival of the city's ethnics) helped keep the demonstrations from raging out of control. Nonetheless, McDermott had felt the impact of the protest while serving in his South Providence command post. The Fourth of July weekend of 1970 was a nightmare. During

The forces of nature can sometimes prove as threatening as individual carelessness or the arsonist's match. Members of the First Unitarian Church on Benefit Street learned that bitter lesson on the evening of August 24, 1966, when lightning struck the spire of the 150-year-old building during a midsummer rainstorm. Ironically, the church had been rebuilt in 1815 after an arsonist destroyed the original structure a year earlier.

The fire caught hold in the organ loft above the nave and quickly spread to the domed ceiling. Firemen began working in the building to extinguish the flames, but they were ordered out by Chief Marshall when fire reached the hundred-foot-high spire. It was feared that the 2,500-pound bell would collapse along with the steeple. The department's aerials were almost fifty feet short of reaching the fire's highest point. Heavy smoke forced firefighters to direct high-pressure hoses from ground positions. By 9:45 P.M. the fire was declared under control, but only after it had caused extensive damage to the interior, which had recently undergone renovation. The church's organ was completely destroyed. Adequate insurance allowed church members to repair an estimated half-million dollars in damage caused by this spectacular blaze. Photograph, August 24, 1966, courtesy of the Providence Fire Department.

Tragedy visited the ranks of the Providence Fire Department on the afternoon of June 29, 1967. Shortly after 1:00 P.M., employees at Mount Pleasant Hardware at 1097 Chalkstone Avenue began to notice smoke rising between the floorboards of the one-story building. A phone call brought three engines and Ladder Truck 6 to the scene. Private Earl T. King, a thirteen-year veteran of the force, jumped from Ladder 6 and headed down the cellar stairs in search of trapped employees. Just before descending into the smoke-filled basement, King was said to have commented, "I hope it's not too bad." With that he disappeared into the darkness.

The flames voraciously fed on paint, turpentine, and roofing materials stored in the cellar. Working with air packs, battering rams, and jack hammers, firemen desperately labored to free their comrade, but they were thrown back by acrid smoke and intense heat. Pressurized cans of paint exploded and shot into the air. A second alarm brought another sixteen pieces of equipment to the scene. Fire Chief James Killilea was felled twice by smoke and was finally taken to Rhode Island Hospital. During the late afternoon a crane was brought in to rip out the northwest corner of the store in an effort to clear an opening to the cellar. Just before midnight a rescue crew located the lifeless body of Private King.

Area businessmen later joined with fire department members in establishing a fifteen thousand dollar trust fund for King's two teen-aged children. Photograph, June 29, 1967, courtesy of the Providence Journal.

James T. Killilea succeeded Lewis Marshall as chief in January 1967. Killilea had a son who was also a Providence firefighter, as well as a first cousin who was serving as chief of the East Providence department when he himself won his promotion. Clearly, firefighting was a family tradition.

Killilea joined the Providence force in 1930. Known in his early days as "Red," the young fireman became wrestling and boxing champion of the police and fire departments in 1931 and again a year later. An equally energetic firefighter, Killilea was injured numerous times in the line of duty and, like Marshall, suffered chronic back problems.

As chief, the sixty-one-year-old Killilea continued the record of excellence set by his predecessor. In 1968 Providence won the Grand Award of the National Fire Prevention Association.

During his three-year tenure Chief Killilea added six new 1,000-gallon-per-minute pumpers, two aerials, and a rescue truck to the department. In 1969 he and Al Mello supervised the overhaul of the fire alarm headquarters, making it one of the most modern fire communications centers in the country. Killilea retired in 1970 and was succeeded in July of that year by John F. McDermott, Jr. He died on December 24, 1982. Photograph, 1966, courtesy of the Providence Fire Department.

one seven-hour period firefighters were summoned to douse 43 building fires, 236 bonfires, and 14 auto fires. This hectic period also included 29 false alarms, 15 rescue runs, and several hydrant openings. The vast majority of such incidents were in South Providence, with Federal Hill a distant second.

In July 1971 the mindless vandalism reoccurred. The *Providence Journal* decried the "savage pattern of physical attacks on firemen responding to calls in some neighborhoods" where "men were showered with rocks and bottles." This national holiday, read the editorial, had "been turned into a macabre orgy of department baiting" that put the city "on the brink of disaster." Alarms on the four-day weekend totalled 1,420. Fortunately for the department and public order, conditions became much less severe in 1972 and succeeding summers as urban unrest subsided throughout the nation. Ironically, Providence won its fourth and final Grand Award in 1971 for its fire prevention effort.

Despite the distractions that surrounded his ascension to office, Chief McDermott lost no time placing his mark upon the fire service. In August 1970 he announced a "streamlining" program that called for the appointment of John F. McDonald, Jr. as assistant chief and the elimination of four battalion chiefs by reducing the number of fire districts from four to three. In the reshuffling the Allens Avenue facility replaced the Point Street station as the seat of Battalion 2, heralding the ultimate abandonment and demolition of that landmark structure (in which both the chief and his father had served) in September 1971.

As a union man, McDermott remembered the rank and file. In March 1971 firemen began working an average week of forty-two hours in four platoons. This six-hour decrease, negotiated by the union, was instituted six months ahead of schedule because of the acquiescence of McDermott and Mayor Doorley.

The major fire of the McDermott era occurred on July 8, 1971, at the American Screw Company complex at Randall Square just before its renovation and rehabilitation was scheduled to begin. A furious three-alarm fire destroyed four vacant mill buildings. Only the strenuous efforts of fourteen engine companies and five ladder units saved the remaining structures for recycling and modern use.

In the early seventies the union expanded its agenda. Under the persistent leadership of Private Edward Tavares and state union officials like Frank Montanaro of Cranston, it pressed for such concessions as the elimination of the residency requirement for firefighting personnel and the establishment of educational incentive programs for firefighters identical to those granted to policemen in 1968. It achieved a major victory in the latter quest in May 1970 when union lobbyists, working in concert with Professor Patrick T. Conley of Providence College, drafted and secured the passage of "An Act Establishing an Incentive Pay Plan for Municipal Firemen" which set up five incentive salary steps, the highest of which—possession of a baccalaureate degree—entitled the firemen to a 16 percent increase above basic salary.

This landmark legislation helped spur an unprecedented and highly desirable educational boom in the fire service that has continued to the present. A 1982 educational inventory by James Mirza for the Local 799 newsletter contended that Providence then had a total of 134 firefighters who had earned either associate or bachelor's degrees, most of the latter from Providence College and most of the former from Roger Williams or the Community College of Rhode Island.

In March 1973 McDermott retired, leaving the chief's post—at least temporarily—to his assistant, John F. McDonald, Jr., of Fourth Street in the Mount Hope section of the city. Like Killilea, the new chief was an accomplished athlete. Having starred in sandlot baseball as a youth, McDonald also served as president of both the baseball and the softball umpires' associations of Rhode Island. His elevation to the department's top post was primarily a tribute for years of able service, similar to that accorded Chief Fischer in 1937, because the sixty-three-year-old appointee revealed his intention to retire in August—five months hence—at the time of his elevation. True to his word, McDonald departed on schedule to make way for Michael Moise, twenty years his junior and the first chief in the history of the department to hold a college degree.

While the department was in transition, a bombshell fell in the form of an efficiency analysis called the Gage-Babcock Report. This study of the fire service, commissioned by Mayor Doorley in 1972, had been prompted in part by union contentions that the department was undermanned and its stations underheated. Doorley had countered in May 1972 by observing that the lot of firemen had vastly improved since he had assumed office in 1965; wages of privates were up from $106 to $175 weekly, hours had been reduced from fifty-six to forty-two, and the service period for eligibility to retire on pension had been cut from twenty-five to twenty years. Citing population decreases proportionately greater than those of any other city in America (248,674 in 1950 down to 179,116 in 1970), the mayor claimed that the police and the fire departments were overstaffed, since the authorized manpower of both had been set in the 1950s before the great exodus to suburbia. He then announced plans for a study by professional consultants of "the deployment of fire stations and manpower in an effort to provide adequate fire protection at the lowest possible cost."

The Gage-Babcock survey was completed in May 1973. The report opened on a positive note by observing that Providence's force "perhaps more than any fire department in the nation...has adapted itself to changing conditions in the community while also recognizing improvements in firefighting equipment and strategy." When it progressed to specifics, however, the findings were less palatable. The 123-page survey alleged that the department was overmanned, stations were in poor condition, the prevention bureau was inadequately staffed, and the training programs were not sufficiently comprehensive. Its proposed solution was to close four stations, eliminate sixty-two jobs, upgrade training and prevention programs, and build a new central station.

Public discussion began in September, with Local 799, PACE (People Acting through Community Effort), and several vocal Federal Hill and South Side councilmen leading the opposition. When Doorley, on the defensive,

called Providence "the most overprotected city in the country," the union's president, George R. Sullivan, called the Gage-Babcock analysis "biased, deficient and inaccurate." Firefighters went door-to-door to discredit the plan, and over six hundred irate citizens attended a November hearing at the St. Xavier High School auditorium to vent their disapproval with critics, contending that fixed public safety personnel-population ratios should not be applied to Providence because "the nature of the city's population and economy as well as the daily influx from suburbia of workers and persons seeking entertainment require higher numbers of police and firemen." Moved by such pressure, City Council President Robert Haxton pronounced the cutback plan dead in February 1974.

In the turbulent year of 1973 the mayor and the firefighters' union also clashed over the administration's failure to accept the results of binding arbitration that had awarded higher raises than "no-dough Joe" thought fiscally prudent. Local 799's reaction was to seek (vainly) a state statute permitting firemen to strike. On this issue, however, the firefighters could not enlist public support.

Another point of controversy was the residency issue. Through revisions in the Cranston Home Rule Charter in 1973, Cranston firemen became the first in the state to gain the right to live beyond their city's borders. This change was sponsored by Charter Review Commission Chairman Patrick Conley at the urging of Frank Montanaro and Cranston union leaders. Fortified by this victory, Montanaro and Providence's Enrico Landi and Edward Tavares lobbied successfully, over Doorley's opposition, to secure the passage of P.L. 1974, Chapter 185, a state statute declaring that "no city or town shall require that an individual reside within the city or town as a condition for appointment or continued employment in its police or fire department." With the exception of its strike demand, the union had registered impressive wins in Providence against a very formidable chief executive.

On July 29, 1973, Michael Moise of Washington Park became the department's new chief. A nineteen-year veteran of the fire service and the son and the brother of firemen, Moise had had extensive administrative experience within the department, as a U.S. Marine Corps

During the decade of the 1960s, the department reverted to earlier (pre-1909) practice and published informative, illustrated annual reports. This 1968 version shows Providence as "top dog"—the winner (for the third of four times) of the Grand Award of the National Fire Protection Association for the best fire safety program of any city in America. Most responsible for this honor was prevention bureau chief John T. McLaughlin. In 1971, when the city won its fourth and final Grand Award, the city's effort was directed by Fire Marshal Warren R. Kirk. Photograph of cover of 1968 annual report, courtesy of the Providence Fire Department.

The Big Bear market on Hoyle Square was a landmark for West End shoppers since 1937. By 1969 the well-known supermarket had been abandoned, another victim of the massive urban flight from the capital city. On February 19 of that year, at 5:00 P.M.—just minutes after city building inspectors had condemned the structure—fire broke out. Furniture stored on the second floor quickly added to the momentum of the inferno. During the three-alarm blaze fourteen pumpers and five ladder trucks from four communities poured tons of water on the four-story building. At 5:42 P.M. the south wall collapsed, landing just a few feet from several firemen who were scampering for their lives. Three firefighters suffered minor injuries. The remains of the structure were leveled to make way for a bank. Photograph, February 19, 1969, courtesy of the Providence Journal.

On October 29, 1970, a spectacular multiple-alarm fire severely damaged old Hope High School on Providence's East Side. This three-story complex had opened its doors in 1898 and had operated as a high school until 1938, when the new Hope High was constructed directly across the street. The old building was being used as a Veterans Administration clinic when finally vacated in 1969.

The incendiary fire broke out at about 2:49 P.M. Firemen arrived to find flames bursting from the roof of the central portion of the structure. Two firefighters were slightly hurt when the roof of the burning building collapsed. The fire spoiled the hopes of both Moses Brown and St. Dunstan's School for acquiring and refurbishing the city-owned building. Shortly after the blaze the venerable old school was demolished. A modern apartment complex now occupies the site. Photograph, October 29, 1970, courtesy of the Providence Journal.

During the late 1960s and early 1970s, the month of July was greeted with dread by most firefighters. The city's increasing number of abandoned buildings became easy prey for summertime vandals seeking a cheap but dangerous thrill. One of the largest July fires victimized the vacant American Screw Company complex at Randall Square on the evening of July 8, 1971. The cluster of factory buildings occupying the southern portion of the square had been slated for renovation as part of a massive plan to revitalize the area.

Fourteen engine companies and five ladder companies fought furiously to contain the three-alarm blaze. The fire left four multistory brick buildings in complete ruin. Photograph, July 8, 1971, courtesy of the Providence Journal.

Cold weather takes its toll on both men and equipment. Firemen here battle a stubborn blaze at Raymond's Deli-Shop at 1243 Chalkstone Avenue on February 23, 1972. During that night the temperature dropped to zero. An aerial ladder called to the scene was locked with ice and disabled. Providence weathered the cold snap fatality-free, but in nearby Massachusetts and Connecticut six persons died in fires on that frigid February night. Photograph, February 23, 1972, courtesy of the Providence Journal.

officer, and as staff assistant since 1971 to Mayor Doorley. Moise was not only the first college graduate but also the first Italian-American to lead the fire service.

Within two weeks of his elevation, Moise selected Richard Rebello, a Fox Pointer and a colonel in the U.S. Army Reserve, as his assistant chief and thirty-seven-year-old Gilbert H. McLaughlin, Jr., from the Reservoir neighborhood as his administrative assistant. This able and personable trio conscientiously guided the department and enhanced its reputation for efficiency and professionalism. Having survived the initial turbulence of 1973, they have navigated the fire service through generally peaceful waters, a course facilitated by Doorley's mayoral successor, Vincent A. Cianci, Jr., a Republican who constantly courted the union and upgraded the department both to gain the political support of firefighters and their families and to enhance the department's status, morale, and performance.

An early innovation of the Moise era related to the process for selecting new members of the fire service. Having worked for Mayor Doorley from 1971 to 1973 as director of the federal Emergency Employment Program, Moise was familiar with affirmative-action guidelines when he assumed his position as chief. Therefore, he promptly instituted a selection process consisting of impartially administered physical, intellectual, and psychological testing that removed discriminatory barriers to the employment of entry-level firefighters.

Before Doorley left office, he and Chief Moise presented the fire force with its largest accession to date—five 1,000-gallon Mack pump trucks and two 100-foot Maxim ladder trucks—purchased with $464,000 in federal revenue-sharing funds. This apparatus was placed in service on May 23, 1974. In addition, Rescue 3 acquired its first "jaws of life" from the American Fire Equipment Company. After an October 1974 demonstration by Lieutenant Donald Gregson and Private George Boragine, Moise hailed the device as "a fantastic life-saving machine" capable of extricating accident victims from cars with amazing rapidity. Doorley and Moise also planned a new fire department training center, as recommended by Gage-Babcock, but the five-acre Fields Point site they selected was rejected by the Port Development Task Force

appointed after Doorley's departure, and the project was scrapped.

The second half of the 1970s was a period of high, eventually double-digit inflation, and the city budget, dependent on an unresponsive property tax, was severely strained. Yet the Cianci administration met most union demands for raises and fringe benefits and, at the urging of former training chief Edmond Marnane, dramatically improved the pension system, especially for those policemen and firefighters who had already retired with what had become an increasingly inadequate stipend. In addition, Mayor Cianci, with the advice and support of Moise and Robert N. Drummond, chief of the new Emergency Medical Division, introduced telemetry units on rescue vehicles to treat stroke and cardiac victims, acquired the most modern breathing equipment (4.5 Scott Air Paks), and renovated the air supply room at Engine No. 5 on Humboldt Avenue. Because the Great Blizzard of 1978 created a severe transportation problem, Chief Moise secured three military ambulances to travel over otherwise impassable streets. Finally, the department purchased substantial quantitites of the most modern and safe firefighting apparel.

Unfortunately, the inflated dollar could be stretched no further—particularly to buy expensive apparatus or to replace retirees. This situation prompted Local 799's president, James D. Bennett, Jr., to confront Cianci and Moise in March 1979 with the charge that the department had "too few firemen and too many outdated trucks." The shift of the Brook Street station's ladder truck to Admiral Street was the spark that ignited the combustibles. The union approved a strike action and censured the chief. The timely return of the repaired truck and a statement from Cianci that "the city plans to buy between $600,000 and $800,000 worth of new equipment next fiscal year" ended the crisis.

The union's presence, so apparent in the Brook Street confrontation, had been enhanced in September 1977 when Local 799 dedicated Firefighters' Memorial Hall on Printery Street as its permanent home and clubhouse. The financial status of union members was also enhanced when Cianci approved a 1980 contract with an 8 percent pay hike, setting base pay for grade

John F. McDonald, Jr., joined the Providence fire service in 1943 at the somewhat advanced age of thirty-four. In 1957 he earned his captain's badge and was assigned to Engine 14 on Atwells Avenue and later to Ladder 2 on Messer Street. His move up the ranks continued in 1965 with his promotion to battalion chief. When John McDermott took over the reins of the department, he appointed McDonald his assistant chief. Upon McDermott's retirement in March 1973, the sixty-four-year-old McDonald was named to the top post in recognition of his thirty years of distinguished service. McDonald remained as chief only five months, retiring on his birthday, August 11. After moving to Lincoln, he died on March 24, 1979, at the age of seventy. Photograph, courtesy of the Providence Fire Department.

John F. McDermott, Jr., inherited an interest in fire service from his father, who had served as captain of the old Engine 22 on Point Street in South Providence. The younger McDermott joined the department in 1947 and was first assigned to Engine 8 in Olneyville. Active in the successful effort to organize a firefighters' union in Providence, he later served as president of Local 799 for four years.

In 1962 McDermott was promoted to captain and given charge of the Point Street engine company that his father had captained years earlier. He then served successively as battalion chief of the first and the second districts before being named department chief in July 1970, succeeding the retiring Jim Killilea.

As top man, McDermott quickly initiated a plan to reallocate the personnel resources of the department. This reorganization included reducing the number of fire districts from four to three, thus eliminating four battalion chiefs.

During McDermott's tenure an outside consulting firm, Gage-Babcock, was retained by the city to examine the department's distribution of men and equipment with an eye towards increased efficiency. The so-called Gage-Babcock Report was presented to Mayor Doorley in May 1973, two months after McDermott's retirement. Its recommendation to close several stations and eliminate four engine and two ladder companies produced a storm of controversy in the autumn of that year.

With McDermott heading the department, Providence in 1971 won the Grand Award from the National Fire Protection Association despite the problems posed to the fire service by racial violence and urban unrest. The former chief, who assumed office at forty-four and retired at a younger age (forty-seven) than any other twentieth-century Providence chief, still resides in the city and is employed as a fire and safety inspector. Photograph, courtesy of the Providence Fire Department.

The careers of many of these 1974 Fire Training Academy graduates were nearly derailed a year earlier by an attempted reorganization of the department. In May 1973 a report commissioned by the city and carried out by the New York consulting firm of Gage-Babcock and Associates recommended a sixty-two-man reduction in the fire department, the elimination of two ladder and four engine companies, and the closing of four stations. The plan also called for the construction of a new headquarters station near the junction of Mount Pleasant and Chalkstone avenues in the Mount Pleasant section of the city.

The proposed changes brought immediate and angry response from firefighters and a number of neighborhood groups. In November six hundred Providence residents jammed a public hearing to denounce the plan unanimously. Providence Local 799 of the International Association of Firefighters issued a detailed twenty-six-page reply that labeled the Gage-Babcock study "biased, deficient and inaccurate." Public protests and petition drives produced the desired effect. In February 1974 the City Council abandoned the plan. Photograph, May 4, 1974, courtesy of the Providence Fire Department.

three firefighters at $289 per week. The contract also raised the annual clothing allowance by $65 to $315 a year, and the force received one extra paid holiday, bringing its total to eleven. For such generosity, Cianci's opponents in the 1982 mayoral election accused the mayor of entering into "sweetheart deals" with the city's unions, but to no avail.

During the mid- and late 1970s, several significant fires occurred. On January 4, 1975, the century-old Wilcox Building at Weybosset and Custom House streets was engulfed by flames. Fortunately, this heavily damaged structure has been renovated, restored, and reclaimed like a phoenix rising from its ashes. On November 27, 1976, the five-story Swarts Building at 87 Weybosset, nearly opposite the Wilcox, was victimized by an even more destructive blaze. The four-alarm Swarts fire consumed the energies of twenty companies and drew fire crews to Providence from as far away as Swansea, Massachusetts. Burning cinders and tongues of flame, driven by high winds, for a time imperiled the entire financial district. In July 1978 an incendiary fire at the Roger Williams Building on Hayes Street caused $350,000 worth of damage to offices and records of the State Department of Education. In April 1979 a yarn warehouse in the CIC complex burned, causing $300,000 in damage.

Much more tragic than these property losses was a series of fires that took a heavy toll of life. The four-year period from November 1976 to November 1980 is unparalleled in the city's history for fire deaths. In the peak year, 1977, nineteen persons perished as a result of fires, a number unequalled before or since. This grim streak began on November 2, 1976, with a rooming house fire at 6 Adelaide Avenue in the Elmwood section that killed four persons. Three months later two children were burned to death in an Aleppo Street apartment house. Then came the most devastating day of all—De-

cember 13, 1977. In the early morning hours a smoky blaze fed by Christmas decorations ignited on the fourth floor of Providence College's Aquinas Hall, a dormitory for women. Though physical damage was relatively slight, ten young ladies eventually perished in this fire, which ranks as Providence's worst ever. The tragedy, witnessed by hundreds of fellow students, plunged the entire college community into mourning and the city into a state of disbelief. On the evening of that unlucky 13th, another major fire, at the American Legion Post on Bellevue Avenue, claimed the life of Lieutenant William J. Moreland, Jr., and seriously injured twenty other firefighters.

The string of multiple-death disasters resumed on January 4, 1979, when three children died in a house fire on Potters Avenue. In October a father and son burned to death on Swan Street, and in March 1980 a family of three perished in Silver Lake when their home at 136 Pocasset Avenue was destroyed. Ending this tragic skein was a Ford Street house fire on November 21, 1980, in which a man and three young children died. Despite all the advances in communications, training, equipment, and techniques, fire and smoke could still be lethal.

One aspect of local firefighting that has assumed great importance in the past twenty years is the problem of arson, its detection and prevention. The anti-arson campaign began in 1966 when Mayor Doorley instituted a program to demolish vacant structures in the Model Cities area of Upper South Providence. By the early seventies dilapidated buildings were being razed throughout the city at the rate of three hundred to five hundred a year. Meanwhile, a police arson squad was created in March 1969 to step up criminal investigations of suspicious fires. During the squad's first eight years of operation, it conducted an average of 224 arson-related investigations annually. In a 1977 report to the police chief, Detective Lieutenant Bernard E. Gannon gave his analysis of the

motives for local arson in the order of their importance: "vandalism, economic gain, revenge, concealment of a crime, pyromania, and intimidation."

After a state anti-arson effort, launched in 1980, collapsed for lack of federal funding, Providence city government stepped up its campaign against this dangerous crime. In January 1983 Mayor Cianci, Chief Moise, and Public Safety Commissioner Sanford Gorodetsky announced the creation of a seven-man Arson Prevention Unit headed by the fire department's Captain John Butler. This new group, working in concert with the staff of Fire Marshal Tom Doyle and neighborhood organizations such as SWAP, has effected a significant decrease in building arsons and has doubled prosecutions for this crime. The unit's initial success helped to earn the city a twenty thousand dollar Ford Foundation grant in early 1984 to set up an early-warning computer system to identify arson-prone buildings.

During the decade of the 1980s, the department has attempted to reconcile economy with efficiency. Approximately $1.3 million was allocated for additional apparatus under an innovative lease-purchase arrangement. From March 1980 through mid-1983, the department acquired a total of ten trucks via this economical procedure. The most notable accession was a special-hazards unit purchased in 1981 because of Moise's concern over the increased use by businesses of hazardous materials and the threat such use poses to firefighters operating at fires or other incidents where such dangerous substances are involved.

In 1982 the chief struck a real bargain when he bought two Mack Aerialscope trucks from New York City's surplus inventory. These units, which have since been refurbished, are each equipped with a "flying" watergun and a telescoping boom that can lift a bucket straight up to a height of seventy-five feet to battle high-rise fires more effectively.

Limitations on city funds have compelled the department to revive the old rather than acquire the new. In the past several years the renovation of existing stations and the thorough repair and refurbishment of existing apparatus has been a continuous and unremitting process. Such renewal has occurred on the most comprehensive scale in the department's history, and at a great saving of tax dollars.

The possession of a college degree, coupled with the availability of a twenty-year pensioned retirement (or less with military service), has prompted more and more firefighters to seek a second career. A notable early example of this trend was John Hawkins, who earned a bachelor's and a master's degree as a member of the fire service (1959-1966), attended law school, and launched a career as an attorney and politician. Hawkins entered the state senate in 1967 and reigned as majority leader from 1973 to 1976. As a legislator he facilitated the passage of several measures of importance to firefighters. Today he is the attorney for Local 799.

Another enterprising firefighter was James Creamer, who became the first member of the fire service to graduate from Providence College while on active duty. Having received his A.B. from the School of Continuing Education in 1972 when he was a private at Engine No. 8, Creamer, an ex-Marine, next obtained a master's degree in education, retired, taught for a time in the Providence school system, and then embarked upon a third career in state government.

Other highly educated firefighters have remained with their initial calling, using their advanced learning to teach other firemen or to improve the administrative efficiency of the department. Among these professionals are Fire Marshal Thomas Doyle, arson squad leader John Butler, and Battalion Chief Donald Gumbley, who teaches courses in fire science at Providence College.

But in Providence officers do not have a monopoly on college degrees. The rank and file also demonstrates a passion for higher education. Preeminent in this respect is Private Richard F. Kless. When he joined the department in 1979, Kless had already earned a B.S. in social

Nineteen seventy-five marked the one hundredth anniversary of the Wilcox Building, an elegant five-story brick structure on Weybosset Street near the corner of Custom House Street. The year began, however, not with celebration but with a conflagration that almost spelled this landmark's doom. On January 4, shortly after 2:30 P.M., fire broke out in a stairwell of the building. Within minutes the blaze had reached the elevator shaft, and by the time firemen arrived, flames were shooting fifty feet above the roof. Employees on the upper floors quickly scrambled to safety. It took the combined efforts of six ladder companies and twice as many engine companies to contain the inferno. As the fire intensified despite an aggressive interior attack, Chief Moise, sensing danger, ordered all firefighters out of the flaming structure just minutes before the center of the third, fourth, and fifth floors collapsed with a roar. Businesses on the lower floors incurred severe water damage. The chief casualty was Dana's Old Corner Book Store, a well-known Downtown fixture for more than sixty years, which suffered a loss of approximately seventy-five thousand used and rare books.

On the following day the Providence Journal announced that the building would probably be demolished, but instead city officials and preservationists combined to restore the century-old landmark to its former elegance. Photograph, January 4, 1975, courtesy of the Providence Journal.

On November 27, 1976, the Wilcox Building was the victim of a second fire. At about 7:00 P.M. firemen were called to a blaze in the five-story Swarts Building on Westminster Street. Located just a few yards from the historic Arcade, the Swarts Building had recently been vacated and was slated for demolition. The four-alarm blaze brought twenty fire companies to the scene. Equipment from thirteen communities in Rhode Island and nearby Massachusetts was called to help contain the flames or man the city's emptied fire stations. Thirty police officers were dispatched to control a crowd of almost five thousand spectators.

Before long, the fire was threatening the five-story Lauderdale Building. Firefighters doused the building while others worked to extinguish flames spreading along walkways connecting the two structures. Chunks of flaming sparks, driven by a westerly wind, descended on businesses and homes as far away as Benefit Street. Soon the roof of the fire-damaged Wilcox Building was ablaze, and to some anxious spectators it appeared as if the entire commercial district might fall prey to the spreading conflagration. But firefighters doused the Wilcox fire and by

10:30 P.M. brought the Swarts blaze under control. Demolition crews moved in the following day to complete the job of razing the charred Swarts Building to make way for a parking lot. Photograph, November 27, 1976, courtesy of the Providence Journal.

work and a master's degree in religious studies (with distinction) from Providence College, where he had co-captained the football team as an undergraduate. Kless then returned to Providence College to acquire his third degree, a B.S. in fire science. Simultaneously he began to teach courses in basic Catholicism and Christology as a visiting lecturer in PC's School of Continuing Education. In 1982 the members of Local 799 recognized his leadership potential by electing Kless their secretary-treasurer. By 1984 the Diocese of Providence had recruited Private Kless to conduct a seminar on "the spirituality of work," wherein he expounded his "firm belief that unions should be advocates for social reform and human progress."

The statistics of today's progressive department give an indication of its present scope and activities. On December 31, 1983, the firefighting force consisted of 479 uniformed and 21 civilian personnel, assigned to fifteen engine companies, eight ladder companies, three rescue units (stationed in a triangular pattern at Allens Avenue, North Main Street, and Hartford Avenue), and four service divisions—fire prevention, training, communications control, and maintenance. During 1983 engine companies responded to 25,793 alarms, ladder units answered 4,756, and rescue squads made 13,994 runs. A total of 909 building fires occurred, 245 of which were incendiary in origin. Most of the 198 vacant-building blazes were the work of arsonists. Rubbish, brush, vehicle, and other miscellaneous fires totalled 3,730. The operating budget for fiscal year 1983-84 ($13.1 million) was money well spent, with Providence continuing to maintain its elite 2A rating classification by the Insurance Services Office—

the highest overall rating awarded to any community in the nation. Leadership was effective, morale high, and Local 799, led by Leo Miller, vigilant and strong.

In the years immediately ahead, one principal change in the department could be in the area of personnel. Despite Moise's affirmative-action selection formula for entry-level positions, the Providence fire force has never included a woman, and so-called minorities—blacks, Hispanics, and Asians—are greatly underrepresented, though they now constitute nearly one-quarter of the city's population. In the 1979 class of recruits, for example, all fifty-seven fire department openings were filled by white males at a time when the city had a black population of eighteen thousand and Hispanic and Asian totals of nine thousand and four thousand respectively. Since the new home rule charter (effective January 1983) contains a residency requirement that supersedes the nonresidency statute of 1974, and since that requirement has been reinforced by a court decision and a specific statute, the future pool of recruits for the highly sought-after job of firefighter should more nearly reflect the city's diverse population.

Whatever the future, the department's past has been bright and its present even more lustrous. Today's Providence firefighter is more highly educated, better trained, better paid, better equipped, more professional, and more highly respected than ever before. Despite its commendable 230-year record of performance and community service, the "good old days" for the Providence Fire Department are now!

Few Providence residents will forget the tragedy that befell the Providence College campus during the early morning hours of December 13, 1977. The Christmas season had begun like any other for the three hundred coeds occupying the four-story U-shaped Aquinas Hall dormitory. In the spirit of the season, many had decorated doors and windows with tinsel and paper decorations. In the northeast corridor of the top floor, a group had constructed a manger display of paper and cardboard that rested on three trashcans. It was illuminated with a gooseneck lamp for an added touch.

At 2:57 A.M. firemen were alerted by the pulling of a box alarm on the fourth floor. Some students failed to take the alarm seriously at first because several false alarms had been pulled earlier that year, but when a heat and smoke detector went off a short time later, the students became aware of the imminent danger.

By the time the first firemen arrived, thick smoke and flames were pouring from several fourth-floor windows. Firemen spotted Christine Manuel and two of her roommates leaning from their windows screaming for help. Despite pleas from firefighters to remain calm and await rescue, two leaped from the window to their death forty feet below. Less than a minute later rescuers snatched Miss Manuel from her smoke-filled room. Some coeds tried to escape down the fiery corridor, while others remained in their rooms, stuffing sheets and blankets around doors to keep out the smoke. On the ground, groups of students were pressed into service assisting firemen by pushing cars from the parking lot to allow fire equipment to maneuver into position.

Thirty-eight minutes after the first alarm was sounded, the fire was declared out. As firemen penetrated the smoke-filled corridor,

they discovered the lifeless bodies of five other students. During the following weeks three other PC coeds succumbed to injuries suffered in the fire, bringing the total to ten—the largest number of fatalities in a single fire in the city's history.

Subsequent investigation revealed that the thirty-eight-year-old building did meet minimum state fire code requirements. The fire was confined to the fourth floor corridor and one room.

During that tragic year nine other Providence residents lost their lives in city fires, including Lieutenant William J. Moreland, who perished in a Bellevue Avenue building fire of incendiary origin just hours after the PC disaster occurred. Photographs, December 13, 1977, courtesy of the Providence Journal.

Other contacts between Providence College and the fire department were highly positive, especially the degree program in fire science initiated by PC in the fall of 1973. As consultants and enthusiastic supporters, Chiefs John McDermott and Michael Moise (PC, class of 1952) worked closely with Dean Roger L. Pearson in designing a curriculum that would complement the existing fire science programs in Rhode Island and southern New England. The thrust of the curriculum was to look beyond the associate's degree to a baccalaureate that would prepare the professional firefighter for a second career in fire science once he retired from active service. To that end, Providence College is the only institution to grant a bachelor of science degree in fire science in southern New England.

The first graduates of the program received their associate's degree in May 1974, and the class of 1977 included seven B.S. degree recipients in fire science. Through 1983 the college had conferred ninety-two associate's degrees and seventy-four bachelor of science degrees in fire science upon members of area fire departments.

This novel program at Providence College, both in its inception and continuance, has been closely tied to the Providence Fire Department, and a large number of Providence firefighters possess Providence College degrees. The relationship is further strengthened by the Reverend

Francis D. Nealy, O.P., chaplain of the Providence department. Father Nealy, a PC professor, has extended his official duties as chaplain to include service as academic advisor to fire science students.

Those primarily responsible for the ongoing success of this collaborative educational effort are shown here: (left to right) Chief Moise, Father Nealy, Dean Pearson, and Chief Don Gumbley, instructor in fire science and administrative assistant to the chief. Photograph, 1983, by Louis Notarianni.

In June 1976 the Providence firefighters' union purchased a two-story brick building on Printery Street for conversion into a union hall. The structure, built in 1886, was originally part of the Allen Printworks complex. During the first decades of the twentieth century the building housed the Scientific Textile Finishing Corporation, the meter-repair department of the Providence Gas Company, the Alrose Chemical Company, and finally Morris Cohen and Sons Furniture. The Providence Firefighters' Memorial Hall was dedicated on September 10, 1977, and still serves as headquarters for Firefighters' Local 799. Photograph by Denise Bastien.

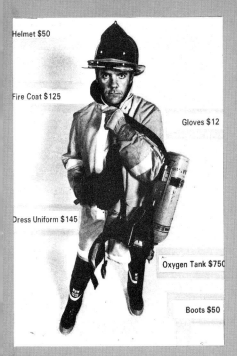

Helmet $50

Fire Coat $125

Gloves $12

Dress Uniform $145

Oxygen Tank $750

Boots $50

This 1979 composite illustration depicts a modern, fully attired Providence fireman. The cost of his apparel exceeds the annual salary of his 1879 predecessor. The upgrading of personal dress and safety equipment has been a major concern of the Moise administration. By 1985 the cost of most of the listed equipment had nearly doubled. Illustration, 1979, courtesy of the Providence Journal.

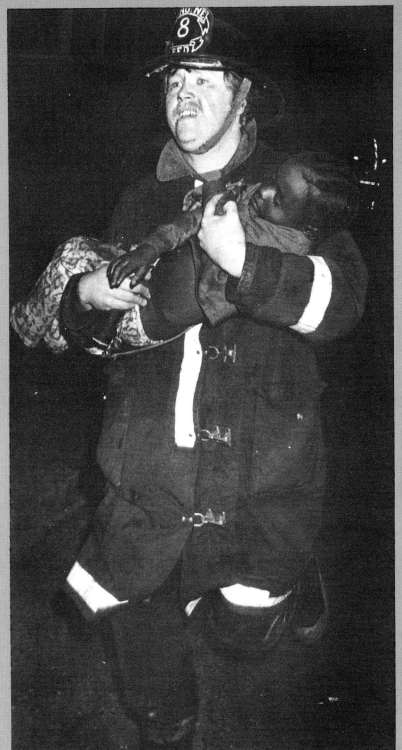

The job of the firefighter is such that he can be transported in minutes from the calmness or conviviality of the station to a turbulent, life-threatening task, or to a heartrending scene like that of mangled bodies in an auto crash or the charred remains of innocents whose lives were taken by smoke or flame while they slept. Here fireman Gary E. Mulcahy carries a dead child from the ruins of a January 1979 Potters Avenue house fire that claimed three young residents. Photograph, 1979, courtesy of the Providence Journal.

The great flight to the suburbs in the 1950s and 1960s accelerated the decline of America's inner cities. Failing downtown businesses and neighborhoods pockmarked with deteriorated or abandoned housing stock resulted in an explosive growth of incendiary fires. By the 1970s federal crime officials had labeled arson the fastest-growing crime in the country.

The city of Providence was not exempt from this scourge. During the first nine months of 1969 the fire department responded to nearly three hundred arson-related fires. In March of that year a two-man arson squad was established in the Providence Police Department. Stepped-up investigation, however, represented only one small element in the battle against this multifaceted crime. A 1977 arson squad report prepared by Detective Lieutenant Bernard Gannon credited the state with having adequate arson laws but cited the immediate need for "stern punitive measures" against arsonists in the Rhode Island courts. A recent in-depth study of arson in Rhode Island conducted by reporter Christopher Scanlan of the Providence Journal-Bulletin substantiated Gannon's charge. That study revealed that between 1978 and 1982, 71 percent of all those convicted of arson in Rhode Island received a suspended or deferred sentence or probation.

In 1980 the state of Rhode Island received a major boost in its war against arson when the federal Law Enforcement Assistance Administration (LEAA) provided nearly $350,000 for

the establishment of a statewide arson unit in the attorney general's office. A prosecutor was hired along with a paralegal, two investigators, two accountants, and a chemist. Among its functions, the unit held seminars for police and fire officials. Concurrently a Governor's Task Force on Arson was formed to study the problem. By 1982, however, a decline in federal funding forced serious cutbacks in this statewide effort.

In Providence the war continued. Pictured here is a victory for the arsonists, an incendiary blaze in April 1981 that caused extensive damage to the historic 122-year-old Trinity United Methodist Church. The fire began in a nearby tenement house at 18 Bridgham Street, injuring six firemen and leaving thirty-five homeless. The Bridgham Street house (above) had been the site of at least three burnings during the previous four months.

Fires such as this caused growing community concern, especially in the city's poorer neighborhoods. That same year representatives from a number of community groups banded together to form the Providence Anti-Arson Coalition. This SWAP-led effort (Stop Wasting Abandoned Property, Inc.) began working with city officials to board up and secure abandoned buildings from vandals. It also helped lobby for 1982 laws requiring absentee owners of multifamily dwellings to reveal their insurance companies and compelling insurers to pay all municipal liens before reimbursing an owner for his loss.

In January 1983 Mayor Vincent Cianci, Jr., announced the creation of a seven-member Arson Prevention Unit composed of fire investigators and police detectives and headed by the fire department's Captain John Butler. This agency, which replaced the two-man police arson squad, has helped to account for a sharp drop in building arsons and a doubling of arson arrests, according to statistics compiled by Butler and Fire Marshal Thomas Doyle.

In 1984 the city was the recipient of a $20,000 Ford Foundation grant to conduct a two-year pilot program to computerize arson data. This project is currently analyzing information about properties in the Elmwood, South Providence, and West End neighborhoods in an effort to identify high arson-risk properties and thus provide preventionists with warning.

But despite all these efforts, the problem persists. A Journal-Bulletin computer study revealed that during the three-year period from January 1981 through December 1984 Providence recorded 4,283 cases of arson, resulting in over $12 million in property damage. Of these cases, building arsons accounted for 1,265 incidents and $11,552,300 in losses. For fire officials and victims of arson, the major finding of the two-year study was not encouraging—"In Rhode Island arson all too often remains a crime that is invisible, ignored and virtually risk free." Photograph, April 4, 1981, courtesy of the Providence Journal.

With the city budget strained in the early 1980s by double-digit inflation, Mayor Cianci and Chief Moise had to resort to creative financing in order to acquire expensive new apparatus. The plan they employed, called lease-purchase, froze the price of a vehicle at its cost when the initial contract was signed and allowed the city to apply its lease or installment payments towards the outright acquisition of the vehicle.

On March 3, 1980, the Board of Contract and Supply approved a lease-purchase contract in the amount of $1.3 million for the acquisition of ten trucks—five triple combination pumpers, one triple combination pumper with a fifty-five-foot articulating boom and nozzle, three tractor-trailer one-hundred-foot aerial ladder trucks, and one "special-hazards" heavy-duty rescue vehicle.

From July 1980 to July 1983 these units were placed in service one by one. The culmination of this acquisition program occurred on July 11, 1983, when the tenth truck was commissioned at the North Main/Meeting Street station as Ladder 4. Mayor Cianci (shown here) was the speaker at these dedication ceremonies, where all ten vehicles were displayed. To his right is Deputy Chief Gil McLaughlin and Chief Moise. Hidden beside McLaughlin is the event's master of ceremonies Sanford Gorodetsky, public safety commissioner from July 1979 to April 1984 and the son of Louis Gorodetsky, a Providence firefighter. Photograph, July 11, 1983, by Barry Greenlaw, courtesy of Vincent A. Cianci, Jr.

Faced with business and industry's increased use of hazardous materials—highly explosive chemicals, substances emitting lethal fumes, and the like—the department in 1981 fashioned its own response: the special-hazards unit. This multipurpose aluminum-body truck, built by the American Modular Body Corporation of Smithfield, Rhode Island, on a Mack R-model chassis, measures twenty-seven feet in length, ten feet eight inches in height, and weighs thirteen tons. The unit is equipped not only to handle crises involving all types of hazardous materials but also to cope with more routine emergencies. Its varied gear includes exposure, acid, proximity, and entry suits; maxi-force air bags; hydraulic jacks; the "jaws of life"; power saws and other assorted cutting equipment, including an oxyacetylene torch; a generator; fire extinguishers; pumps; smoke ejectors; positive-pressure breathing apparatus; and 4.5 Scott cylinders. According to Chief Moise, the unit and its crew are intended to function both as a specialty squad for extraordinary emergencies and as a manpower squad to provide backup service at regular fires.

To allow firefighters to provide for their own safety as well as that of the general public, Eugene Patrone's Division of Training formulated a hazardous-materials program, initially designed and taught by Donald Gumbley and Ronald Raymond. Its purpose is to show firemen how to identify such substances and how to initiate actions at "hazmat" emergencies. Photograph, 1981, courtesy of the Providence Fire Department.

Among the newest pieces of equipment added to the department's fleet are these two Mack Aerialscopes. The trucks feature a telescoping boom that can lift a bucket seventy-five feet or direct a water gun from that height. The bucket is also capable of rescuing as many as eight people from a burning multistory building.

The budget-conscious fire chief purchased the eleven-year-old vehicles from the city of New York for $50,000 and spent an additional $117,000 refurbishing them. Purchased new, the tower ladders would have cost nearly $700,000. The trucks, presented to the city on August 27, 1982, were assigned to the LaSalle Square and Messer Street stations. Photograph, 1982, courtesy of the Providence Fire Department.

Several old fire stations in Providence have been converted to new uses. All the fire stations pictured on these pages were closed during the department's modernization program from 1949 to 1951.

The old Water Witch, or "Sixes," station on Benevolent Street (bottom right) was sold at auction in March 1951 to Bryant College, who renovated its façade and converted it for use as a library. Later Brown University acquired the building and uses it to house its student radio station WBRU. Both Hose Company 2 (bottom left) on South Main Street and Hook and Ladder 7 (top, far right) on the corner of Hope and Olney streets were converted into restaurants. In the 1970s the South Main Street structure opened as a nightspot bearing the name "Engine Company No. 2." It has since undergone several changes in use and occupancy. The Hope Street station, now Tortilla Flats and Domino's Pizza, has remained a familiar landmark on the East Side.

The old station house of Niagara Steam Engine Company 5 (center, far right) is now shared by a computer retailer and an architectural design studio. Still visible above the North Main Street entrance is the old insignia of the Niagara Fire Company.

Ladder Company No. 5 on Public Street at Burnside (bottom, far right) was transformed into an employment center in the late 1960s during the War on Poverty. When federal funds ran out, it was closed and then vandalized. In the early 1990s, it was destroyed by an arsonist. Photographs by Denise Bastien and Louis Notarianni.

The Outlet Company, founded in 1894, became Rhode Island's largest department store and flourished until the 1960s, when suburban malls and urban exodus combined to effect its rapid decline. In October 1982 this Weybosset Street landmark closed, dealing a severe blow to the city's retail shopping district.

On October 16, 1986, as downtown developers prepared to purchase the sprawling five-story structure for conversion to apartments and retail

erupted at daybreak on the fifth floor, burst through the roof, and shot fifty feet into a slate-gray sky, where the smoke generated was visible from Newport. A hundred firefighters battled to save the doomed landmark, but at day's end the Outlet was a smoldering hulk. The rapid and effective response of Providence's firefighters saved the outer walls and lower floors and gave rise to hopes that the structure could be salvaged, but in the end, given the building's location on a

course of action. Arson was cited as the probable cause of Downtown's most spectacular fire in decades.

Neighboring Johnson and Wales University eventually bought the site to expand its urban campus. By 1994 an imposing college dormitory and the beautifully designed Gaebe Commons had risen phoenixlike from the Outlet's ashes. Photograph, October 16, 1986, by William L. Rooney, courtesy of the Providence Journal.

chapter FIVE

Epilogue, 1985-2001

Because sixteen years have elapsed from the initial publication of this volume, and it has been fifteen years since its third and last printing, a reissue of *Firefighters and Fires in Providence* for the 2002 dedication of Providence's new Public Safety Complex demands at least a brief update.

Since 1984, when the narrative ended, Providence and its fire service experienced significant change and development. Joseph R. Paolino, Jr., assumed the mayoralty in April 1984, when Mayor Vincent A. Cianci, Jr., was forced to resign. Paolino first served as acting mayor by virtue of his council presidency, and then as chief executive in his own right after winning the popular election in November. His six-and-one-half-year tenure ended in January 1991 after he declined another term in favor of a run for governor—a race he lost.

Paolino, a Democrat, initially maintained a cordial relationship with the firefighters' union, Local 799, and enlisted former firefighter and union activist James Creamer as one of his top aides. Creamer served on loan from the office of Matthew Smith, Speaker of the House and a close Paolino associate. In the late 1980s, however, sparks began to fly during contract talks between the union and the city administration.

Rising above partisanship, Chief Michael Moise continued to lead the department until 1990, when health problems influenced his decision to retire after seventeen years at the helm. As his successor, Paolino appointed Moise's friend and deputy Gilbert H. McLaughlin. The much beloved and highly respected Moise died in March 1997.

During the Paolino-Moise era the five most significant fire-related developments were (1) the construction of a modern communications building in 1985 on West Exchange Street to house Alfred J. Mello's Department of Communications, which included the fire department's Bureau of Operational Control; (2) the spectacular Outlet Company fire of October 1986, which destroyed the vacant downtown building that had housed the state's largest department store; (3) the devastating blaze in Olneyville's Riverside Mill complex in December 1989, described by Chief Moise as "the worst fire during my administration in terms of area or scope"; (4) the expansion of the Emergency Medical Division in 1988 by the addition of two rescue units—Rescue 4 at LaSalle Square and Rescue 5 at the North Main Street Station; and (5) the adoption of the Incident Command System, a procedure developed by the California forest-fire corps, whereby the battalion chief at a major fire sets up a defined command center that assigns specific duties to firefighters on the scene.

The spectacular and suspicious fire at the Riverside Mill symbolized the plight of Providence as an arson-prone city. When the FBI released its annual crime statistics in 1990, Providence ranked third (up from fourth) in the nation for arson, based on the number of incidents (449) reported per 100,000 population. In concert with the office of the state fire marshal, the department began a vigorous antiarson program that eventually lessened the frequency of this crime, but Providence was an old industrial city in a postindustrial age, with scores of abandoned mills and factories. In addition, it was much more densely populated than newer cities in other parts of America. In 1990 its mere 18.91 square miles contained a population of approximately 160,000 (down from 267,918 in 1925), and its neighborhoods harbored hundreds of vacant three-decker houses on tiny (3,200 square-foot) lots. Providence was a firebug's paradise.

In July 1990 Gilbert McLaughlin, a thirty-three-year veteran, called by Paolino "a true professional," assumed command of the department in a City Hall ceremony that featured the newly formed fire department color guard and a bagpipe rendition of "Amazing Grace" by firefighter Richard Hughes. On September 30 of the following year McLaughlin abruptly announced his retirement, citing the worsening of a job-related injury to his neck and shoulders that he had suffered years before. By this time Vincent "Buddy" Cianci had reclaimed the mayoralty (which he still holds). Both Cianci and union president Stephen Day expressed surprise and regret over McLaughlin's early departure—a testimony to the chief's ability and popularity.

Although McLaughlin initiated several significant programs during his brief tenure, especially one to combat arson, the most notable innovation on his watch came in June 1991, when Cianci and the chief announced that the 1991 Fire Training

In Providence the Communications Department is one of three coequal divisions of the Department of Public Safety, along with the Police and Fire Departments (see the organizational chart in Appendix IV). As a practical matter, however, one of the main functions of Communications is to serve as the Fire Department's Bureau of Operational Control.

In November 1985 the city dedicated a new $2.7 million state-of-the-art communications building, hailed as "the most modern dispatching facility in the Northeast," on West Exchange Street. When this strategic center opened, five three-man teams, operating in eight-hour shifts, handled five telephone lines, a network of 1,370 alarm boxes, several computers, radios, and tape recorders, and an electronic strategy map of the city to keep constant track of all engines, ladders, rescue vehicles, specialty units, and personnel, and to dispatch them promptly and effectively.

This building—the central dispatch for police as well as firefighters—has been dedicated to Alfred J. Mello, the city's first director of communications. Chief Mello (inset) progressed from a civilian employee of the fire department assigned to the alarm division (1949), to firefighter alarm technician (1954), to radio engineer with the rank of lieutenant (1958), to the department's fourth superintendent of fire alarm (1965), in which capacity he supervised the design and installation of a new central fire alarm system for the city.

In 1971 Mello resigned from the fire service to accept appointment by Mayor Joseph Doorley as the first director of the new Department of Communications, a safety division charged with the responsibility of coordinating telecommunications for all municipal agencies. In 1987, two years after presiding over the transfer of the fire alarm system from the old Kinsley Avenue facility to the building that now bears his name—a move carried out without allowing a single called alarm to go unanswered—Mello retired. When Mayor Vincent A. Cianci, Jr., returned to office in 1991, he prevailed upon Mello to resume his old post for one year to reorganize the Department of Communications. Mello, who had become one of Cianci's most trusted advisers, complied, completed his task, and retired a second time in January 1992, after which the West Exchange Street building was named in his honor. Photograph, 2002, by Peter Goldberg; inset photo, courtesy of Alfred Mello.

Academy class would include seven women applicants among a total of sixty students. In January 1992, during the tenure of its new chief, Alfred F. Bertoncini, the Providence Fire Department swore in its first two female firefighters. The pioneering duo were Heidi Verity, twenty-five, of Harmony, a former Smithfield firefighter, and Melissa Talbot, twenty-two, of Pawtucket, a former prison guard. The eminently quotable Cianci, calling their appointment "long overdue," observed that it is tougher to get accepted and get through the Providence Fire Academy than it is to get accepted to an Ivy League college today.

Chief McLaughlin's deputy and Training Academy classmate Alfred F. Bertoncini was Cianci's "obvious choice" to assume direction of the department in October 1991, and he was formally sworn in two months later. A thirty-five-year veteran of the force, he brought not only skill and experience to the position but also a lifelong desire for firefighting dating back to his childhood visits to the departmental carpenter shop and fire station on Manton Avenue near his family home.

Bertoncini led the department for over three years before retiring in December 1994. As the former assistant drillmaster at the Training Academy and an active union member, he was especially concerned to implement standards to ensure the health and safety of his firefighters.

Chief R. Michael DiMascolo, Bertoncini's successor, was the most prominent union member to attain the rank of chief. A longtime labor activist, DiMascolo had been elected vice president of Local 799 and served diligently in that capacity for three years prior to his elevation to the rank of deputy assistant chief and fire marshal for the city in December 1993. DiMascolo became head of the department on December 18, 1994. Although he retired as chief six months later at age forty-seven before he could leave a permanent imprint on the force, DiMascolo continued his firefighting career and presently holds the responsible post of chief deputy state fire marshal.

In the mid-1990s the city considered the possibility of closing two fire stations—Rochambeau and LaSalle Square. This proposed plan was part of a citywide analysis examining basic operations of city government with a view towards increased efficiency. This was partly a response to the severe recession that had affected the state and the city following the banking crisis of 1990–91. A consultant was hired to examine these issues. When word of the analysis leaked out, East Side residents descended on City Hall, armed with petitions with hundreds of signatures. Also, the firefighters union, led by Stephen Day, peppered the neighborhoods with flyers, circulated thousands of S.O.S. buttons ("Save Our Stations"), and attended public meetings with so-called "Towering Inferno" videos. In the end, the experts hired by the city, MMA Consulting Group, Inc., of Boston made a number of recommendations regarding organizational improvements in the department, but recommended against closing stations.

Although the last decade and a half has been marked by progress in the fire service, it was also marked by pathos. Two of Providence's worst fires in terms of lives lost occurred during that span. The first, which was also the city's most deadly arson, occurred on a frigid February evening in 1993 on Hayward Street, extinguishing the lives of an immigrant family of six. The other, on Hymer Street in the Wanskuck neighborhood, killed Deborah Griffin and four of her children in their third-floor apartment and injured nine others, two seriously. The December 8, 2000, fire struck as the thirty-five-year-old Mrs. Griffin was completing her final year of studies at Rhode Island College en route to a career as a music teacher in the Providence school system.

Crises sometimes create heroes. So it was on Hymer Street. Before firefighters were summoned, two passers-by, Dara Veng and Jean Melo, braved dense smoke and heat to rescue two young brothers from the first-floor apartment. After Veng and Melo carried Michael and Nelson Andino to safety, they tried in vain to reach the Griffin family on the third floor. Six nights later Michael W. Green rescued three children from a house fire on Erastus Street in the Manton section that claimed the life of an infant boy. Eventually Veng, Melo, and Green were awarded prestigious Carnegie Medals by the Andrew Carnegie Hero Fund Commission, a Pittsburgh-based foundation that honors American heroes who risk their lives to save others. Their exploits prove that a vigilant and supportive citizenry can be the firefighters' best ally.

The pervasive and increasing influence of Fire Fighters Local 799 of the International Brotherhood of Fire Fighters is another aspect of our story. Under the leadership of past president Stephen T. Day and current head William Farrell, the union has safeguarded the welfare of the fireman in such varied areas as pay, promotions, minimum manning, equipment replacement, fire prevention, job safety and security, education, and pensions. Day, the superintendent of the department's automotive division, hails from a family of avid firefighters; his father, Frank, was a lieutenant, his brother Michael is a battalion chief, and two other brothers, Peter and the late Timothy, also served with distinction on the force and in the union. George Farrell, the union's president since 1996, is a third-generation firefighter. As chairman of the Rhode Island Fire Safety Board of Appeal, he participated in drafting a new rehabilitation code for Downtown commercial buildings that could allow their reuse for loft living units.

Several significant changes in state law are the

From 1989 through 2001, several major fires destroyed old mills, most of which were vacant. The most notable mill blazes—all caused by arson—included the cafeteria building at the Gorham Manufacturing complex (March 1991), the American Tubing and Webbing Company on Gordon Avenue (October 1996), the Steere Mill on Wild Street in the Wanskuck section (August 2000), the former Louttit Laundry on Cranston Street (May 2001), and a three-story building on Bosworth Street in Manton, which originally housed a branch of the Centredale Worsted Mills (June 2001).

Most spectacular of all, however, was the conflagration depicted here which destroyed the 126-year-old Riverside Worsted Mill complex on Aleppo Street in Olneyville in December 1989. Providence's first general-alarm fire in twenty years, the blaze was fought by 159 men. Unlike the mills mentioned above, the twelve Riverside buildings were occupied, housing a diverse array of some ninety tenants that included jewelry shops, a paint store, two theater groups, an automotive shop, a plastics company, and numerous artists' studios. A Providence Journal headline summarized the fire's impact: "Artists and Artisans See Flames Devour Their Life's Work."

As tragic as the Riverside fire was, however, it paled in comparison to a December 1999 fire at a vacant warehouse in Worcester, Massachusetts, only forty miles distant from Providence, that took the lives of six Worcester firefighters. The massive funeral ceremonies staged to honor their bravery and mourn their deaths were attended by dozens of their Providence comrades. Photograph, December 19, 1989, by Bob Breidenbach, courtesy of the Providence Journal.

Certainly the department's most notable and significant recent development has been the emergence of women as full-fledged firefighters. This trend began in June 1991, when Mayor Cianci and Chief McLaughlin announced the selection of seven women for the Training Academy's forty-second class. The first two female cadets from that historic class—Heidi Verity and Melissa Talbot—were sworn in by Cianci and Chief Bertoncini on January 15, 1992. Both Public Safety Commissioner John J. Partington and the ceremony's principal speaker, former Superior Court presiding justice Anthony Giannini (whose son William is now deputy chief), noted the significance of this event for all sixty-four cadets, and especially for the path-breaking female pioneers.

This photo depicts recruit Verity rappelling the seven-story tower at the Training Academy on Dexter Street under the watchful eye of her drill-master, Lieutenant John Thomas.

During the past decade several other women have chosen firefighting as a career and survived the rigorous screening and training process to become members of the force. Presently (in 2002) there are ten female firefighters serving in various capacities throughout the department.

Though successful, the debut of the woman firefighter has not always been smooth. Some have suffered crude insults and ostracism from their male counterparts similar to that endured by African Americans during the early episodes of integration. One woman fought back vigorously. Julia O'Rourke, who was subjected to repeated sexual harassment from several firemen, sued the city in 1995 and won two civil rights trials against it. In July 2001 the U.S. District Court ordered the city to pay firefighter O'Rourke damages in the amount of $551,818

for the harassment inflicted upon her. Although she took a medical leave from her job because of stress, O'Rourke courageously returned to work in the department's Fire Prevention Bureau. There are three women enrolled in the upcoming 2002 training class. Photograph, January 15, 1992, by Mary Murphy, courtesy of the Providence Journal.

Chief Gilbert H. McLaughlin began his service in August 1957 as a trainee. His first assignment was as a hoseman with Engine Company No.3 on Franklin Street. McLaughlin's administrative talents were recognized by his superiors after his promotion to lieutenant in 1969, and he was assigned to headquarters as an aide to Chief McDermott in 1970. The following year he was elevated to the rank of captain and given the prestigious command of Ladder No.1 at LaSalle Square. During the long tenure of Michael Moise, McLaughlin served as the chief's administrative assistant and right-hand man, and having continued his rise in rank to deputy chief, he became Moise's logical and deserving successor in July 1990. Mayor Paolino, who appointed him, cited not only McLaughlin's administrative and planning experience but also his two citations for "heroic action" and his several citations for meritorious action.

Like several chiefs before him, McLaughlin was plagued by job-related injuries which prompted his retirement on October 23, 1991, at the age of fifty-five. Despite his brief tenure, McLaughlin has been credited with starting a comprehensive antiarson program, establishing systematic replacement of fire apparatus while changing truck color to red and white, and launching renovations to fire stations across the city. He also helped to activate a computerized system for identifying hazardous materials, and he helped establish a mobile command post for hazardous materials incidents and major fires. Chief McLaughlin increased the force by sixty members in accordance with successful union demands for minimum manning, and he presided as well over the enrollment of the first females into the department's Fire Training Academy. Photograph, courtesy of Gilbert H. McLaughlin.

132

Chief Alfred F. Bertoncini, a native of the Olneyville-Manton section of Providence, grew up looking forward to a career in fire-fighting. After serving four years in the U.S. Air Force (1953-1957), which he joined while the Korean War was still in progress, he enrolled in the eighteenth Training Academy class and secured appointment to the force on August 5, 1957. Cited three times for meritorious service, Bertoncini rose through the ranks to become deputy assistant chief in November 1982 and assistant chief in October 1990. He directed operations at such major blazes as those at the Outlet Company (October 1986), the Union Station (April 1987), and the Riverside Mill complex (December 1989).

At the time of his formal elevation to chief in December 1991, Bertoncini was described by Deputy Assistant Chief Enrico Landi, his close friend and 1957 classmate, as a man who knew the job from both sides—union and administration. "He has negotiated for the union, and he has negotiated for the city," said Landi, "and no matter what side he was nego-tiating for, Al had the respect of both sides."

As chief, Bertoncini made the health and safety of his men a primary concern and regarded the implementation of the National Fire Protection Association's Standard No. 1500, relating to occupational safety and health, a major step towards that end. He retired on December 18, 1994, after more than thirty-seven years of service, leaving his firefighter son David to carry on a family tra-dition. Photograph, courtesy of Alfred Bertoncini.

result of union lobbying efforts and/or similar cam-paigns of information and persuasion conducted by the Rhode Island Fire Chiefs' Association and the Office of the State Fire Marshal. In 1986 the legisla-ture made the installation of home and business smoke/fire detectors by sellers mandatory, and in 2001 it passed the nation's first statute requiring the residential installation by sellers of carbon monoxide detectors. In Providence the marshals of the Fire Prevention Bureau conduct the inspections to certify proper compliance with these safety laws. Another key measure enacted by the General Assembly at the urging of the firefighting fraternity was the 1987 adop-tion into state law of NFPA Standard 1500, which set legal requirements for Fire Department Occupational Safety and Health.

Since June 1995 the Providence Fire Department has been led by Chief James F. Rattigan of Fox Point, the son of firefighter John A. Rattigan, Sr., whose career spanned thirty-two years. Chief Rattigan assumed his present post at age forty-three, joining Mike Moise as the youngest head of the department. During Rattigan's tenure the fire service has made sig-nificant technological advances: full computerization, portable radio equipment for all mobile positions, the establishment of fire ground-radio frequencies, and the acquisition of state-of-the-art firefighting appara-tus, including a systematic program of replacement.

Chief Rattigan's most tangible and enduring lega-cy, however, may well be his important role in the construction of the new Providence Public Safety Facility, which he and his staff will occupy in 2002. This elegant, yet functional, facility, hovering over I-95, will give the Providence Fire Department public visibility and recognition commensurate with its meritorious history.

The firefighting force, dominated in the nine-teenth century by old-stock Rhode Islanders, called "Yankees," and in the twentieth by men of Irish or Italian ancestry, will undoubtedly experience another change in complexion during the century ahead. In the 2000 federal census, Providence regained its tra-ditional position as New England's second largest city because of a huge influx of Hispanic arrivals. Presently, 30 percent of Providence's population (52,146 out of a total of 173,618) have Latin American roots. The city's ever-changing demographic is reflect-ed in the composition of the two most recent training classes—both the forty-fifth and forty-sixth schools have approximately one-third minority representa-tion, and most of these recruits are Hispanic candi-dates for the fire service. Department personnel may change, but its tradition of courageous, selfless serv-ice remains immutable.

The worst city fire in terms of lives lost since the Providence College tragedy of December 1977 occurred at 54 Hayward Street in South Providence during the early morning hours of February 27, 1993. Shortly after midnight, in a misdirected and misguided act of revenge, two teenagers tossed a gasoline-filled "Molotov cocktail" into the front stairwell of the three-family house. While the occupants of the first and second floors scurried to safety in the icy and frigid night, the family of Carlos and Hilda Chang became trapped in their third-floor apartment. Hilda's brother, Ivan Ponce, was severely burned and suffered other injuries when he jumped from a window, but the six Changs—husband, wife, and four children—perished in the flames. Two rescues and nine fire companies responded to the fire, which took more than two hours to bring under control.

Carlos Chang, who was partly of Chinese ancestry, had migrated to Providence from Guatemala in 1990 and secured work at the Point Judith Fishermen's Coop to earn money to bring his family to Rhode Island. His wife and four children arrived in 1991 and became residents of the ill-fated Hayward Street tenement.

The deaths shocked Providence's small, close-knit Guatemalan community, in which Carlos was an active participant. The fate of the Changs was made even more tragic by the senseless nature of their death. The arson was committed

by two youths to avenge an injury done to the cousin of one of the perpetrators in a minor auto accident on Hayward Street earlier in the evening. The Changs were not involved in that incident, nor did they know their murderers. Justice fortunately prevailed. Both perpetrators were caught and tried for multiple murder and arson. One received a life sentence after pleading guilty; the other was sentenced to life in prison without parole subsequent to his conviction for the largest multiple murder in state history. Photograph, February 27, 1993, by Steve Szydlowski, courtesy of the Providence Journal.

All the fires detailed thus far in this volume have been destructive; but regulated fire can be beneficial and even essential. It can also become an aesthetically stimulating art form. In 1995 Barnaby Evans created WaterFire, an award-winning fire display staged upon Providence's newly uncovered rivers and in its recently constructed Waterplace Park. Evans has marshaled hundreds of volunteers to help him present this spectacle on selected evenings in the spring, summer, and fall. The river fires, set in large pots, are fueled by reclaimed pine and cedar wood. They begin at sunset with a backdrop of ethereal music selected by Evans from a variety of sources and cultures. Volunteers in boats tend and replenish the flames until past midnight under the watchful eye of the fire service.

For spectators, fire—controlled and showcased on the water—is an enchanting sight. Since its first performance on December 31, 1994, WaterFire has become a symbol of Providence, the Renaissance City. Photograph, courtesy of Barnaby Evans.

Though Mike DiMascolo's tenure as chief spanned only six months—from December 18, 1994, until his retirement on June 22, 1995—his firefighting career has spanned nearly three decades, and has not yet run its course.

After graduating from Mount Pleasant High in 1967, Chief DiMascolo served on active duty in the U.S. Army for two years during the Vietnam War. Upon his return to Providence he earned an associate's degree from CCRI and a B.S. from Providence College. As his academic studies progressed, Mike won acceptance into the department's thirty-fourth Training Academy on October 29, 1973. After completing intensive training, he embarked upon a series of assignments—at Engine 9, Engine 8, Ladder 2, and as an aide to the chief of Battalion 2—that earned him promotion to the rank of lieutenant in September 1987 and three separate commendations for meritorious service and heroic action.

In August 1991 DiMascolo became a training officer and drill instructor at the Training Academy, and only a year later he was promoted to the rank of captain and assigned to the Fire Prevention/Arson Investigation unit. His meteoric rise continued in December 1993 when he was elevated to deputy chief and fire marshal, and culminated just one year later when Mayor Cianci appointed him to succeed Chief Al Bertoncini.

Chief DiMascolo described himself as "an avid union member." He was more identified with Local 799 and the union movement than any previous chief, having served the union as a member of its executive board and as its vice president for three years. He took special pride in the various charitable projects he directed under union auspices.

Chief DiMascolo's retirement from the Providence force did not end his career. Unlike most of his predecessors, Mike continued as a firefighter. In 1996 he became director of the Rhode Island Fire Academy, and in 1997 he assumed the post of chief deputy state fire marshal under Marshal Irving J. Owens. In that capacity Mike is now responsible for directing twenty-one employees in six separate units: fire investigation, inspections, technical services, plan review, the state Fire Academy, and the clerical staff. Photograph, courtesy of Michael DiMascolo.

A native of the Fox Point neighborhood, the present chief, James F. Rattigan, was educated in the Providence school system, LaSalle Academy, and CCRI before obtaining a B.A. from the University of Rhode Island in 1979. Continuing a family tradition begun by his father, John A. Rattigan, Sr. (whose firefighting career spanned thirty-two years), and followed by his older brother, Lieutenant John A. Rattigan, Jr., the future chief entered the department in 1973. He accepted assignments in various capacities throughout the city, including a thirteen-year assignment at the Messer Street Station in the West End.

In March 1988 Chief Rattigan began his ascent of the organizational ladder with a promotion to lieutenant. He served with the Special Hazards Division and Ladder Companies 7 and 8 (Branch Avenue and Brook Street respectively) through 1994, earning his third meritorious action citation for his efforts at the Hayward Street fire. He rose to the ranks of captain and fire marshal/deputy assistant chief in rapid succession. Finally, in June 1995, Mayor Cianci named Rattigan, aged forty-three, the chief of the department—a position he now holds.

During his tenure Chief Rattigan has overseen the purchase and implementation of state-of-the-art firefighting apparatus and equipment and collaborated with the Department of Communications to obtain portable radio equipment for all mobile positions and to establish fire-related ground-radio frequencies. He has also supervised the computerization of his entire department and reorganized the command staff so that high-level personnel oversee the department's daily operations. Rattigan is a member of the Rhode Island, the New England, and the International Associations of Fire Chiefs. Photograph, courtesy of James Rattigan.

Mike Bates and Paul Doughty

The horrific events on Tuesday, September 11, 2001, marked the beginning of a new war in America—the War on Terrorism. The destruction of the World Trade Center in New York brought into clear focus the need to train fire department personnel in responding to devastation caused by weapons of mass destruction and biochemical ter-rorism. Three years prior to the New York and Washington terrorist attacks, the city of Providence joined a Department of Defense effort to create a Domestic Preparedness Program. Through its "Train the Trainer Program," twenty Providence firefighters received instruction and the department acquired specialized equipment to neutralize the effects of biochemical attacks. Assistant Chief William Gianinni gained national recognition for his expertise in disaster response to terrorist attacks.

This expertise was put to the test within hours of the New York Trade Center tragedy. Three Providence firefighters who had been trained in urban search and rescue—Paul Doughty (right), Mike Bates (left), and Greg Crawford—were called by the Federal Emergency Management Agency (FEMA) and responded immediately. In addition, Providence Battalion Chief Michael Blackburn was one of four fire-fighters from around the nation selected to set up an emergency Critical Incident Stress Management headquarters at Ground Zero. The city provided surgical equipment, water, masks, and a variety of other much-needed supplies. Local blood centers were overwhelmed with poten-tial donors. In the aftermath, Mayor Cianci moved quickly to establish a Public Security Task Force, chaired by Public Safety Commissioner John Partington, to "examine vul-nerabilities within Providence, analyze existing response plans," and devise further protective measures to safeguard the residents of the capital city. The group's report is scheduled to be com-pleted by spring 2002. Photograph, courtesy of the Providence Monthly Magazine (November 2001).

The new Providence Public Safety Facility is nearing completion as this book goes to press. The volume's reissue is inspired by this imposing $60 million project, financed through a bond issue administered by the Providence Redevelopment Agency. Built to replace the cramped and antiquated Fountain Street headquarters, the new structure (shown here) is located just west of I-95 and looks eastward, with a commanding view of the interstate and Providence's Downtown.

Built by developer Vincent J. Mesolella, Jr., contractor O. Ahlborg & Sons, Inc., and the architectural firm of Jung/Brannen Associates, the complex will include police and fire headquarters, the First Battalion fire station, offices for the commissioner of public safety, the municipal court, a public auditorium, and a separate parking garage to accommodate over five hundred vehicles. Its site is bounded by Dean, Washington, and Cottage streets and fronts on Service Road No. 7.

At the groundbreaking ceremony on June 19, 2000, Mayor Cianci stressed not only the build-ing's need and utility but also its aesthetic appeal and its beneficial impact on the run-down

area that was selected for its location. As the building rose, so did morale among police and firefighters. John J. Partington, the city's longest-serving commissioner of public safety, called the project "the lightening rod we need" to generate pride and enthusiasm, and Fire Chief James F. Rattigan expressed the belief that "it will have an effect from the junior man riding a truck to the chief of the department." Providence and its fire service have moved with efficiency and style into the twenty-first century! Photograph, March 2002, courtesy of Peter Goldberg.

136

Bibliographical Essay

The most useful and up-to-date general histories of Providence that provide a backdrop for the development of the fire service are Patrick T. Conley and Paul R. Campbell, *Providence: A Pictorial History* (1982), and John Hutchins Cady, *The Civic and Architectural Development of Providence, 1636-1950* (1957). Older surveys of value for the period they cover include Richard M. Bayles (ed.), *History of Providence County*, 2 vols. (1891); Welcome Arnold Greene, *The Providence Plantations for 250 Years* (1886); and William R. Staples, *Annals of the Town of Providence* (1843). George C. Wilson, "Town and City Government in Providence," doctoral dissertation, Brown University (1889), and Howard K. Stokes, *The Finances and Administration of Providence, 1636-1901* (1903), are specialized, scholarly studies of merit.

A surprising number of local firefighters and fire buffs have attempted surveys of the Providence department, including Elisha Dyer, *Sketch and Reminiscences of the Providence Fire Department* (1886); Captain Charles E. White, *The Providence Fireman* (1886), a detailed, book-length effort; Captain Charles E. White (ed.), *Fire Service in Providence 1754-1904"* (1905), a lengthy unpublished typescript located at the Rhode Island Historical Society; and Charles E. Lincoln, "Providence Fire Department Antedates Revolution," *Board of Trade Journal*, December 1916, pp. 786-88.

More specialized accounts, useful for detail or illustrations, are William H. Mason, *Souvenir Offering of "Fire King Threes" of the Providence Fire Department* (1902 and 1903 editions), and Reuben DeM. Weekes, "Firefighting Equipment in Providence," *American City*, XV (August 1916), 122-26.

General histories of American firefighting were of considerable utility in defining trends and explaining technological innovations. Those we relied upon for this study were Donald J. Cannon, *Heritage of Flames: The Illustrated History of Early American Firefighting* (1977); Paul C. Ditzel, *Fire Engines, Firefighters: The Men, Equipment, and Machines from Colonial Days to the Present* (1976); John V. Morris, *Fires and Firefighters* (1955); Robert V. Masters, *Going to Blazes* (1950); Robert S. Holzman, *The Romance of Firefighting* (1956); John Kenlon, *Fires and Fire Fighters* (1913); and Paul Lyons, *Fire in America* (1976).

More specialized accounts of interest include Elsie Lowe and David E. Robinson, *Bicentennial History of New England Firefighting* (1975); William D. Brinckloe, *The Volunteer Fire Company* (1934); Donald M O'Brien, *"A Century of Progress through Service": The Centennial History of the International Association of Fire Chiefs, 1873-1973* (1973); and George J. Richardson, *Symbol of Action: A History of the International Association of Firefighters, AFL-CIO-CLC* (1974).

Several histories of the fire insurance industry and catalogs of company collections yielded both information and illustrations, particularly John Bainbridge, *Biography of an Idea: The Story of Mutual Fire and Casualty Insurance* (1952), containing a good essay on Zachariah Allen; William G. Roelker and Clarkson A. Collins, *One Hundred and Fifty Years of Providence Washington Insurance Company, 1799-1949* (1949); Frederick T. Moses, *Firemen of Industry: The Hundredth Anniversary of Firemen's Mutual Insurance Company, 1854-1954* (1954); and M. J. McCos-

ker (comp.), *The Historical Collection of the Insurance Company of North America* (1945, 2nd ed. 1967). Helpful technical studies are Phil DaCosta, *100 Years of America's Firefighting Apparatus* (1964); George A. Daly and John J. Robrecht, *An Illustrated Handbook of Fire Apparatus with Emphasis on 19th Century American Pieces* (1972); William T. King, *History of the American Steam Fire Engine* (1896); Rebecca Zurier and A. Pierce Bounds, *The American Firehouse: An Architectural and Social History* (1982); and B. W. Kuvshinoff (comp.), *Fire Sciences Dictionary* (1977).

In addition to the secondary accounts listed above, the serious researcher can avail himself of a number of published or manuscript primary sources. Our study relied primarily upon these.

For the volunteer era, the "Records of the Presidents and Firewards of the Town of Providence, 1805-1852," housed in the City Archives (PCA), is indispensable. Also containing some pertinent facts are the "Providence Town Papers," deposited at the Rhode Island Historical Society Library (RIHSL), and the Providence Town Meeting Records, under the care of the city clerk. The *Providence City Documents* (1854-1908) contain the detailed annual reports of the department, our most valuable source for the era covered. The digests of state laws published at generational intervals since 1719 yielded the text of much fire-related legislation, as did the periodic compilations of Providence city ordinances. For the twentieth century, the annually published ordinances and resolutions of the City Council were also helpful. Specialized manuscript collections at the RIHSL include the "Log Book of Ladder Co. 2" (1867-1874); the "Water Witch Papers"; "Mutual Fire Society Records (1816-1834)"; and the "Records of Hydraulion No. 1 (1828-1853) and No. 2 (1830-1850)." The PCA contains the "Records of the City Council Committee on the Fire Department" for the years from 1896 through 1940, but they are skimpy. The unpublished recent records of the department and its annual reports from 1953 to the present are not well organized. We found them scattered among three depositories: fire headquarters at LaSalle Square, the training center on Reservoir Avenue, and the storage facility on Chad Brown Street. Specialized reports of value include John Ihlder, *The Houses of Providence: A Study of Present Conditions and Tendencies* (1916), a work that sparked zoning and building code reform; Providence Governmental Research Bureau, *Survey and Report on the Providence Fire Department* (1934); and Gage-Babcock and Associates, *Report on Study of Fire Protection: City of Providence, Rhode Island, May 1973* (1973).

Convenient annual statistical data on the department is contained in the *Providence Journal-Bulletin Almanac* (published annually since 1887), the *Providence City Directories*, and the *Providence Pocket Manual* (1913-1964). The two most important newspaper sources, both for facts and illustrations, are the *Providence Journal* (1829 to the present) and the *Board of Trade Journal*, also known as the *Journal of Commerce* and the *Providence Magazine*, 45 vols. (1889-1935). For recent developments, *Special Signal*, the union newsletter, was most enlightening; for pre-1829 incidents we consulted the appropriate issues of the *Providence Gazette* (est. 1762) or the *Manufacturers' and Farmers' Journal*.

APPENDIX I
NOTABLE FIRE DISASTERS IN PROVIDENCE

(An asterisk indicates those fires illustrated with caption in the text)

This listing is selective and subjective. Inflation and urban growth have rendered damage amounts nearly impossible to compare. For example, the Great Blaze of 1801, which inflicted $300,000 in property damage, destroyed nearly 10 percent of Providence's taxable property. For a present-day fire to have a similar impact on the assessed valuation of the city, it would have to consume approximately $140,000,000 worth of property. Obviously, such comparisons are meaningless. In addition, since 1970, estimates of loss are not given for many blazes. The fires we have selected, therefore, were the biggest and most destructive of their day. Some blazes where property loss was slight but loss of life occurred are also listed, especially if those disasters took the lives of firefighters. Human loss is always tragic.

*Eighteen outlying houses burned by Indians, June 28, 1675, eight days after the outbreak of King Philip's War.

*Over two dozen houses burned in the compact part of the town by Narragansetts led by Chief Canonchet, March 29, 1676.

Colony House, Meeting Street, December 24, 1758; $3,600.

I *Waterfront, thirty-seven buildings from Planet to Coin streets, January 21, 1801; $300,000. This "Great Fire of 1801" destroyed nearly 10 percent of the town's taxable property.

*First Congregational Church and other buildings, Benefit and Benevolent streets, June 13, 1814; $40,000.

First Universalist Church and twenty other buildings, Westminster and Union streets, May 23, 1825; $42,000.

South Main Street fire, March 20, 1828; volunteer fireman Joshua Weaver killed, leading to a movement to establish a firemen's relief association in October 1829.

Four-story building (hardware and paint store), South Main Street, December 16, 1837; $20,000.

Dorrance Street Theater, October 25, 1844; $35,000.

Fuller's Machine Shop, Point Street, December 1, 1846; $50,000.

*Anna Jenkins House, Benefit and John streets, November 20, 1849; historic mansion destroyed; two killed.

Tallman and Bucklin's Planing Works, Dyer Street, September 4, 1850; nearly $50,000.

Mill Street fire, Cleveland's Turning Establishment and fourteen other buildings burned, August 5, 1851; $25,000.

Richmond Street Free Congregational Church and adjoining buildings, October 13, 1851; $12,000.

Almy's Waste House, Canal Street, October 23, 1852; $11,000.

Building on Eddy Street, February 13, 1853; $13,800.

Gile's factory, Atwells Avenue, May 13, 1853; $14,000.

Whitaker and Sons store, North Main Street, September 4, 1853; $12,000.

Arnold's Block, North Main Street, October 11, 1853; $50,500; riot erupts, leading to dissolution of the volunteer department.

Howard Block, Westminster Street, October 26, 1853; $240,000.

Lumber yard, Fox Point, October 1, 1854; $15,900.

Roger Williams Free Baptist Church, Burgess Street, January 5, 1855; $13,000.

India Rubber Works, Dorrance Street, April 30, 1856; $13,000.

Dean Steam Planing Works, Dorrance Street, January 11, 1857; $30,000.

Grace Church, corner of Westminster and Mathewson streets, March 14, 1857; $7,000.

Hayward's India Rubber Works and Hope Iron Foundry, five buildings at corner of Clifford and Eddy streets, October 9, 1857; $50,000.

Howard Block, Dorrance Street, November 15, 1858; $177,850.

Pike's Lumber Yard, South Water Street, November 2, 1859; $15,000.

Providence Dyeing, Bleaching, and Calendering Company, Sabin Street, December 6, 1860; $34,400.

Hope Iron Foundry, Eddy Street, February 26, 1864; $44,500.

Seekell's Hollow (Seekell Street near Pine), about thirty buildings, September 22, 1864; $44,500.

Matthewson and Allen's Buildilng, Middle Street, December 31, 1865; $20,000.

Valley Worster Mills, Eagle Street, February 2, 1866; $200,000.

Providence Rubber Works, Dorrance Street, February 9, 1866; $25,000.

Adams and Clafin's Comb Works and other buildings, Pine Street, March 27, 1867; $19,300.

Press Office, Dyer Street, December 31, 1868; thirty-two persons escape via chain from fourth floor.

Lester's Tea Store and other buildings, Westminster Street, August 13, 1869; $24,000.

Tucker, Swan, and Company and Clark's Coal Yard, Dorrance and Dyer streets, June 28, 1870; $69,000.

East Street fire, September 20, 1870; boiler of Steam Engine No. 6 explodes, fatally injuring a bystander and contributing to the death of fireman John H. McLane.

Providence and Stonington Railroad Company roundhouse, Canal Street, October 2, 1870; $50,000.

Mowry and Steere's Lumber Yard, South Main Street, November 7, 1870; $41,000.

Dunnell Block, Canal Street, April 28, 1873; $27,500.

Phoenix Building, Westminster Street, June 28, 1873; $59,000.

Providence Iron Works, India Street, July 21, 1873; $23,000.

Allen's Print Works, North Main Street, February 3, 1874; $75,000.

"Silk Mill," Olneyville, February 23, 1876; $25,000.

Freeman Francis's Stable, Pine Street, May 28, 1876; $35,000.

*C. W. Jenckes and Brother, Harkness Court, Custom House and Pine streets, September 27, 1877; $500,000.

Rose and Eddy Building, Custom House Street, April 13, 1878; $25,000.

*Calendar Street fire, November 21, 1882; four persons killed.

Fletcher's Mill, Valley Street, February 17, 1883; $32,000.

Vaughan Block, Custom House Street, January 18, 1884; $66,000.

St. Paul's Methodist Episcopal Church, Swan and Plain streets, December 5, 1886; hoseman Nicholas B. Duff killed, three other firefighters injured.

*Aldrich House, Washington Street, February 15-16, 1888; $400,000.

*Theatre Comique, Weybosset Street, February 18, 1888; $100,000.

*Daniels and Cornell, Custom House Street, February 19, 1889; $175,000.

*Providence Coal Company, Dyer Street, February 3, 1889; recall sounded May 7, 1889; $50,000.

Shepard Company, Westminster Street, December 5, 1890, $200,000.

*J. B. Barnaby Company, Westminster and Dorrance streets, December 13, 1890; nearly $400,000.

Aborn Street fire, May 21, 1893; $125,000.

*Union Station, February 20, 1896; estimates range from $50,000 to $75,000.

*Masonic Temple, Dorrance Street, March 19, 1896; $260,000.

Flint and Company, Weybosset Street, December 2, 1897; $125,000; one life lost.

Star Theater, formerly the old Normal School, Benefit and Waterman streets, February 11, 1899; $40,000.

Paris House, corner of Eddy and Westminster streets, February 23, 1900; $100,000.

Fitzgerald Building, Friendship and Eddy streets, July 6, 1900; $100,000.

Elmwood Car Barn of the Union Railroad Company, Earl, Bucklin, and Redwing streets, February 18, 1901; $175,000.

Joseph Richards Furniture Store, Dorrance Street, July 16, 1902; hoseman Joseph Devine killed.

*American Ship Windlass Company, Waterman and East River streets, April 22, 1904; $100,000.

*Anthony and Cowell's, Weybosset Street, April 30, 1904; $330,000.

Music Hall and Public Market, Westminster Street, March 16, 1905; nearly $60,000.

Cherry and Webb, Westminster Street, December 7, 1906; $60,000.

Smith Coal Company, South Water Street, January 21, 1907; Captain George H. Noon of Hose 4 fatally injured.

Clark Manufacturing Company, Ashburton Street, April 16, 1907; $20,000.

Tanner's Starch Factory, South Water Street, February 12, 1908; five men killed by explosion; $50,000.

Clark Manufacturing Company, Ashburton Street, December 26, 1908; $25,000; ladderman Benjamin N. Brown killed.

Anthony and Cowell's Furniture Store, Weybosset Street, June 11, 1909; $70,000.

Harbor Junction, Allens Avenue and Kay Street, July 10, 1909; barge *Harrison* loaded with fuel oil; $20,000.

Union Hardware Company, Exchange Place, February 15, 1912; $150,000.

L. Diamond and Sons, Inc., Westminster Street, April 9, 1912; $200,000.

*Revere Rubber Company, Valley Street, May 7, 1912; $500,000; two firemen killed: Lieutenant Christopher Carpenter and Harry H. Howe.

Harbor Junction, Allens Avenue and Kay Street, March 8, 1916; $85,000.

Scattergood Company, North Main Street, April 15, 1916; $50,000.

Elmwood fire, June 24, 1917; two laundries and thirteen houses burned.

Old Park Brewery (Eastern Films Company), August 23, 1917; $100,000.

American Ship Windlass Company, East River and Waterman streets, January 19, 1918; $40,000.

Hurd Brothers'Hay and Grain Warehouse, Dyer Street, July 10, 1919; $53,500.

Foster-Smith Company, garage, Mathewson Street, July 28, 1919; $42,000.

Providence Arcade, Westminster and Weybosset streets, October 29, 1919; $200,000.

Slade Mansion, Elmwood Avenue, December 25, 1919; $41,500.

William J. Gilmore Garage, 57-59 Grant Street, January 8, 1920; twenty-four automobiles destroyed.

Scattergood Block, 179 Canal Street, January 15, 1920; $300,000.

Hoppin Homestead Office Building, 357-359 Westminster Street, February 16, 1920; $50,000.

Lorraine Hotel (formerly Newman House), Aborn Street, February 18, 1920; $110,000; three killed.

American Enamel Company, Neville Street, July 2, 1920; $165,000.

2 * Washington Bowling Alleys, Washington Street, January 31-February 1, 1921; four firemen killed; Lieutenant Michael J. Kiernan, Thomas H. Kelleher, John J. Tague, and Arthur Cooper.

Webster Wool Noils and Waste Company, Lester Street, May 25, 1921; $60,000.

Furlong Apartments, North Main Street, February 13, 1923; $75,000.

*Shepard Company, Westminster Street, March 8, 1923; $1,500,000.

Morris and Cummings Dredging Company, dredge No. 5, in harbor, September 17, 1923; $200,000.

Waterfront coal and grain district, Dyer Street, October 9, 1923; $500,000.

A. T. Scattergood Company, furniture store, 110 North Main Street, February 22, 1924; $200,000; five firemen hurt.

Times Block, Westminster and Stokes streets, February 25, 1924; $45,000; seven firemen hurt.

American Enamel Company, Neville Street, May 3, 1924; $100,000.

Phillips-Baker Rubber Company, Warren Street, May 25, 1929; $250,000.

St. Joseph's Hospital, Broad Street, February 23, 1930; $89,800.

*State Pier, Allens Avenue, February 25, 1931; nearly $500,000.

Joseph Marcus and Company, North Main Street, January 1, 1934; $50,000; four firemen hurt.

New England Grocery, 93 Weybosset Street, May 23, 1934; $200,000.

Andrews and Spelman, hay and grain, Globe and Eddy streets, June 3, 1934; $75,000.

Weybosset and Union streets, eleven stores gutted, March 2, 1936; $100,000.

G. D. Del Rossi, macaroni factory, India Street, July 3, 1938; $75,000.

Conrad Building, Westminster Street, four alarms, December 3, 1938; $100,000.

Scott Furriers, Inc., Westminster Street, January 28, 1939; $100,000.

Eastern Coal Company, Allens Avenue, July 6, 1939; $100,000.

McCarthy's Freight Terminal, Allens Avenue, October 7, 1939; $100,000.

Point Street Grammar School, January 4,
1940; $500,000; ten injured, including six firemen.

*Union Station freight office building, February 18, 1941; $150,000; six firemen injured.

*Infantry Hall, South Main Street, October 4, 1942; $160,000; two firemen hurt.

Holy Rosary Church, Fox Point, October 5, 1942; severely damaged by fire set by arsonist Donald Bennett, an escapee from the Exeter School for the Feeble-Minded, who also set the Infantry Hall fire on the previous day.

*Rheem Shipyard, Field's Point, December
3 31, 1942; $1,700,000.

Charles H. Robinson, Inc., Leonard Street, July 15, 1943; $100,000.

F. H. Buffinton Company, Eddy Street, December 15, 1944; $100,000.

Cleinman and Sons Jewelry Company (old Federal Street School), Federal and Barker streets, July 4, 1946; $50,000; seven firemen injured.

Curran and Burton Coal Company wharf, Allens Avenue, September 21, 1946; $220,000.

Warehouse Realty Corporation, Union Avenue, January 23, 1946; $250,000.

American Waste Paper Company, 24 Ash Street, October 1, 1946; $150,000.

*Rhode Island Recreation Center, North Main Street, November 13, 1947; $650,000.

Industrial Building, Crary Street, February 28, 1949; $150,000.

Terminal Warehouse Company of Rhode Island, Inc., Allens Avenue, July 7, 1949; $150,000.

Brick warehouse building, 179-189 Canal Street, February 23, 1950; $300,000.

Low Wholesale Company Building, 393 Harris Avenue, July 2, 1952; $100,000.

Swift and Company, meat distributing plant, 252-258 Canal Street, January 8, 1953; $100,000.

Taco Heaters, Inc., Elmwood Avenue, April 29, 1953; $100,000.

L. H. Tillinghast Plumbing Supply Company Building and four additional buildings in area at 162 Dorrance Street, September 22, 1953; $300,000.

Unoccupied building, 32-38 Snow Street, January 22, 1953; $313,000.

Three-story Plymouth Building, 114 Mathewson Street, February 5, 1956; $126,000.

United Camera-Franklin Appliance, 603-619 Westminster Street, February 5, 1956; $216,000.

Leavitt-Colson Company, 47 Pine Street, November 4, 1956; $201,000.

United Plumbing & Heating Supply Company, 237 West Exchange Street, November 30, 1956; four firemen injured.

Classical High School, May 6, 1958; $150,000.

School administration building, Pond and Summer streets, July 25, 1958; $600,000.

Lyra Brown Nickerson Settlement House, main building, Olneyville, February 20, 1959; $125,000.

Lodging house, 11 Wilson Street, February 1, 1962; six persons died; $32,000.

Abandoned house, 6 West Park Street, February 20, 1963; Lieutenant Joseph F. Dorsey killed.

*Two oil tank trucks collide, Washington Bridge, April 18, 1963.

*Foremost Rubber Company, 245 West Exchange Street, October 18, 1965; $62,000.

Mercantile Building, 221-239 North Main Street, May 22, 1966; $143,000.

*First Unitarian Church, 285 Benefit Street, August 23, 1966; $75,000.

Shipyard Drive-In Theater, 1 Washington Avenue, Providence-Cranston, September 4, 1966; $200,000.

Federal Dairy Company, 83 Greenwich Street, January 25, 1967; $200,000.

Eddy and Fisher Wholesale Liquors, 387 Charles Street, June 27, 1967; $452,000.

*Mount Pleasant Hardware Store, 1097 Chalkstone Avenue, June 29, 1967; $238,000; Private Earl T. King killed.

Mount Pleasant Dairy, 33 Dearborn Street, June 24, 1968; $140,000.

El Rio Café and Artie's Grill, 402 Broad Street at Trinity Square, July 24, 1968; $100,000.

St. Michael's Church Parish Hall, Prairie Avenue, December 31, 1968; $270,000.

Jefferson Apartments, 288 Broad Street, February 6, 1969; $203,000.

*Former Big Bear market and two buildings, Hoyle Square, February 19, 1969; $68,500.

*Old Hope High School, 331 Hope Street, October 29, 1970.

*American Screw Company complex, Randall Square, July 8, 1971; largest fire in several years destroys vacant historic mill complex.

Jefferson Apartments, 288 Broad Street, June 12 and August 5, 1972.

Lederer Building, 100 Stewart Street, June 4, 1973.

National Collapsible Tube Company, 362 Carpenter Street, August 26, 1973.

*Wilcox Building, Weybosset and Custom House streets, January 3, 1975.

Rooming house, 6 Adelaide Avenue, November 2, 1976; four persons killed.

*Swarts Building, 87 Weybosset Street, November 27, 1976, a four-alarm blaze fought by twenty fire companies.

Apartment house, 55-57 Aleppo Street, February 7, 1977; two children killed.

4 *Aquinas Hall, dormitory for women, Providence College, December 13, 1977; ten women died.

American Legion Post, 126 Bellevue Avenue, December 13, 1977; Lieutenant William J. Moreland, Jr., killed and twenty firefighters injured.

Family Court and State Education Department offices, Roger Williams Building, Hayes Street, July 21, 1967; $350,000.

*Dwelling house, Potters Avenue, January 4, 1979; three children killed.

Yarn warehouse, CIC complex, April 25, 1979; $300,000.

Apartment house, 303-305 Swan Street, October 17, 1979; father and son killed.

House, 136 Pocasset Avenue, March 4, 1980; family of three die.

Railroad Freight House No. 2, 312 Canal Street, September 12, 1980; historic Thomas A. Tefft-designed structure destroyed.

House, 151 Ford Street, November 21, 1980; man and three children die.

*Six buildings, including Trinity United Methodist Church, Trinity Square, April 4, 1981; six firemen hurt.

Mill building at Olneyville Square, 34 and 38 Dike Street, August 7, 1981; man killed, three firefighters injured.

Tenement house, 440 Plainfield Street, February 10, 1982; woman killed, three firefighters injured.

Apartment building, 12 Bodell Avenue, March 5, 1982; man killed.

Three-family house, 4 Courtland Street, March 21, 1982; woman killed.

Boarded-up Lexington Avenue School, August 19, 1982; one firefighter injured.

Apartment house, 53 Wesleyan Avenue, August 19, 1982; woman killed.

Former Academy Avenue School, September 8, 1982.

Apartment house, 57 Brownell Street, December 2, 1982; two killed.

House, 54 Pocasset Avenue, December 25, 1982; man killed, two firefighters and one woman hurt.

Schubert Heat Training Company, 282 Richmond Street, November 30, 1983; one killed, seven injured.

*Former Outlet Company Department Store on Weybosset Street severely damaged by arson fire, October 16, 1986.

Old Union Train Station suffers severe damage in arson fire, April 27, 1987.

*Riverside Mill Complex in Olneyville destroyed in spectacular fire, December 18, 1989.

Cafeteria building in vacant Gorham Mill Complex destroyed by arson, March 20, 1991.

The Walsh Athletic Center at Rhode Island College burns, January 5, 1992.

*Six members of the Chang family perish in an arson fire on Hayward Street, February 27, 1993. The two perpetrators receive life sentences for multiple murder.

The Shepard Building, a Downtown landmark, severely damaged by arson fire, June 11, 1994.

Firefighter John McKenna nearly dies after being trapped in the basement of a 79 Detroit Avenue house fire on January 29, 1996, then retires from the force following a long hospitalization.

Two children die in a house fire at 21 Atlantic Avenue, February 6, 1996.

One killed and four injured during a fire at 264-66 Manton Avenue, February 21, 1996.

Arson fire destroys a five-story vacant Gordon Avenue Mill in South Providence, October 22, 1996.

House fire on Pine Street kills one man and seriously injures a woman, June 2, 1997.

Fire seriously damages the Providence Produce Market on Harris Avenue, November 23, 1998.

Eleven Brown University students flee a third-floor fire at 44 Brook Street during the early morning hours of March 10, 1999.

Vacant Steere mill on Wild Street in Wanskuck destroyed, August 19, 2000.

A woman and her five-year-old son rescued from their second-floor apartment during a fire at 98 Pocasset Avenue, September 29, 2000.

Firefighter Arlindo Rodrigues injured when he falls through a roof while extinguishing a fire at an industrial building at 7 Ricom Way, October 6, 2000.

Hymer Street fire in Wanskuck kills a mother and four of her children, December 8, 2000. Two other children are seriously burned. Their civilian rescuers, Dara Veng and Jean Melo, later receive Carnegie Medal for Heroism.

Infant dies as flames engulf a two-story house at 88 Erastus Street on the morning of December 14, 2000. Three other children saved by heroic act of neighbor Michael Green, who receives the Carnegie Medal for Heroism. The victim is the eighth child to die of fire-related causes in Rhode Island during an eleven-day period.

Blaze engulfs two buildings at 1422-1428 Broad Street on March 2, 2001, leaving five families homeless and a resident severely injured after jumping from a third-floor window.

Former Louttit Laundry Building on Cranston Street, a historic landmark, partially destroyed by arson, May 13, 2001.

Large three-story vacant industrial building on Bosworth Street in Manton that originally housed a branch of the Centredale Worsted Mills destroyed by arson, June 16, 2001.

Two house fires on June 27, 2001, one on Waverly Street and the other on Kossuth Street, displace thirty-four people.

Fire destroys convent of the Dominican Sisters of the Presentation at 157 Congress Avenue, Elmwood, December 19, 2001.

1 Greatest relative fire loss.
2 Highest toll in death and injury to firefighters.
3 Highest dollar-amount fire loss.
4 Largest fire death toll.

APPENDIX II

Chief Engineers and Chiefs
of the
Providence Fire Department

Volunteer Era

Henry G. Mumford (1838)
Smith Bosworth (1838-1839)
Allen Peck (1839-1840)

Henry G. Mumford (1840-1841)
Joseph W. Taylor (1853-1854)

Paid Force

Joseph W. Taylor (1854-1859)
Thomas Aldrich (1859-1862)
Charles H. Dunham (1862-1865)
Dexter Gorton (1865-1869)
Oliver E. Greene (1869-1884)
George A. Steere (1884-1909)
Holden O. Hill (1909)
Reuben DeM. Weekes (1909-1921)
William F. Smith (1921-1923)
Frank Charlesworth (1923-1937)
John H. Fischer (1937)

Thomas H. Cotter (1937-1951)
Lewis A. Marshall (1951-1967)
James T. Killilea (1967-1970)
John F. McDermott, Jr. (1970-1973)
John F. McDonald (1973)
Michael F. Moise (1973-1990)
Gilbert F. McLaughlin (1990-1991)
Alfred F. Bertoncini (1991-1994)
R. Michael DiMascolo (1994-1995)
James F. Rattigan (1995-   )

APPENDIX III

Legally Constituted Governing Agencies
of the
Providence Fire Department

Board of Fire Commissioners
(Appointed by the City Council)

Stillman White	Feb. 27, 1895, to Mar. 3, 1902
William H. Luther	Feb. 27, 1895, to Dec. 19, 1901
Dexter Gorton	Feb. 27, 1895, to Jan. 4, 1904
Benjamin F. Harrington	Jan. 6, 1902, to Jan. 4, 1904
Ira Winsor	Mar. 3, 1902, to Jan. 4, 1909
William Andrews	Jan. 4, 1904, to Mar. 25, 1911
James Davis	Jan. 4, 1904, to July 4, 1905
George L. Greene	Jan. 1, 1906, to Jan. 5, 1915
Arthur H. Smith	Jan. 4, 1909, to Jan. 5, 1914
George Hunt	Apr. 10, 1911, to Feb. 7, 1916
John R. Dennis	Feb. 2, 1914, to Oct. 6, 1924
Ralph S. Hamilton, Jr.	Feb. 1, 1915, to Jan. 7, 1918
William H. Covell, Jr.	Feb. 7, 1915, to Feb. 3, 1917
Everett J. Horton	Feb. 21, 1917, to Feb. 6, 1928
Clarence M. Dunbar	Jan. 7, 1918, to Feb. 7, 1927
Howard D. Wilcox	Oct. 7, 1924, to Jan. 4, 1926
Walter J. Gilbert	Jan. 4, 1926, to Apr. 15, 1931
Elmer S. Cowan	Feb. 7, 1927, to Apr. 15, 1931
William G. Thurber	Feb. 6, 1928, to Feb. 2, 1931

Board of Public Safety
(Appointed by the governor with the advice and consent of the senate)

Benjamin P. Moulton	Apr. 15, 1931, to Feb. 1, 1932
George T. Marsh	Apr. 15, 1931, to Jan. 2, 1935
Michael H. Corrigan	Apr. 15, 1931, to Jan. 2, 1935
Everitte St. John Chaffee	Feb. 1, 1932, to Jan. 2, 1935

Bureau of Police and Fire
(Appointed by the mayor subject to the approval of the City Council)

Thomas H. Roberts	June 8, 1935, to Jan. 3, 1938
Benjamin P. Moulton	June 8, 1935, to Jan. 1, 1940
Joseph C. Scuncio	June 8, 1935, to Jan. 2, 1939
	Jan. 6, 1941, to Apr. 18, 1951
Francis J. O'Brien	Jan. 3, 1938, to Jan. 6, 1941
Howard S. Almy	Jan. 2, 1939, to Jan. 16, 1942
Edward L. Casey	Feb. 6, 1941, to Apr. 18, 1951
William H. Garrahan	Jan. 16, 1942, to Nov. 1, 1945
Benjamin M. McLyman	Nov. 1, 1945, to Apr. 18, 1951

Commissioners of Public Safety
(Appointed by the mayor subject to the approval of the City Council)

John B. Dunn	Apr. 19, 1951, to Sept. 10, 1959
Francis A. Lennon	Oct. 13. 1959, to Dec. 31, 1964
Harry Goldstein	Jan. 1, 1965, to Apr. 14, 1973
Francis B. Brown	Apr. 25, 1973, to Feb. 25, 1974
Leo P. Trambukis	Sept. 9, 1976, to July 7, 1979
Sanford H. Gorodetsky	July 7, 1979, to Apr. 24, 1984
Charles A. Pisaturo	Jan. 1, 1985, to Jan. 2, 1990
John Partington	Jan. 2, 1990 to

Note: Prior to the implementation of the strong-mayor city charter in January 1941, these agencies exercised de facto as well as de jure control over the personnel and policies of the fire department. Since 1941, especially under Roberts, Doorley, and Cianci, the mayor, acting in concert with the fire chief, has been the dominant influence. In fact, the mayor acts as public safety commissioner when vacancy in that office occurs.

Organizational Chart of the Providence Fire Department, 2001

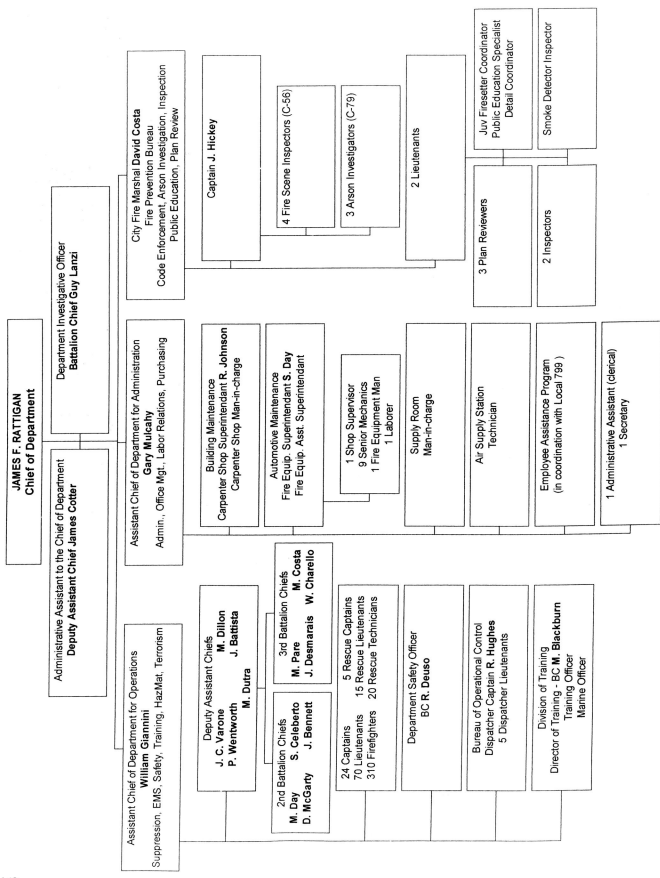

APPENDIX V
Schedule of Apparatus and Equipment, 2001

Apparatus	Fuel	Average Age	Output	Capacity	Special Technical Ability
15 Frontline Engines	Diesel	Six Years	1,250 gpm*	500-750 gal.	One of these Engines is equipped with ability to dispense foam and water
Eight (8) Frontline Ladders	Diesel	Eight Years	N/A	N/A	1 Aerial Truck – 100' 5 Aerial Trucks – 110' 1 Tower Ladder – 75' 1 Tower Ladder – 95'
Special Hazards	Diesel	Nine Years	N/A	N/A	Contains Jaws of Life, cutting torches, air bags, HazMat spill containment, gas meters, etc.
Five (5) Rescue Trucks	Diesel	Five Years	N/A	N/A	All are equipped with Advanced Life Support Units
One (1) Air Supply Unit	Gasoline	10 Years	N/A	N/A	Mobile Unit able to dispense air packs, cylinder and other support equip
One (1) Foam Tender	Diesel	20+ Years	N/A	N/A	Supplement to above described Engine Company
Marine I (Fire Boat)	Gasoline	10 Years	500 gpm	N/A	In service nine months of year; Patrols Narraganset Bay and Port of Prov.
Three (3) Water Rescue Crafts	Gasoline	Unknown	N/A	N/A	Boats utilized on an "as needed" basis and towed to scene by apparatus

THE DEPARTMENT OWNS AN ASSORTMENT OF TOWABLE TRAILERS OF VARYING SIZES FOR THE STORAGE AND DELIVERY OF MISCELLANEOUS EQUIPMENT, AS FOLLOWS:

Dive Trailer	Dive team equipment, scuba gear, a variety of communications equipment, rope, and markers.
Decontamination Trailer	Equipment necessary for the decontamination of mass casualty victims. Carries portable showers, tents, heaters, backboards, etc.
HazMat Trailer	HazMat Entry Suits, tents, heaters, decontamination equipment, and materials designed to stop leaking containers.
Collapse Trailer	Assorted shoring equipment, lumber, tools, and other equipment used in the removal of trapped victims from collapsed structures.

* gallons per minute

The average age of Front Line apparatus is: 6.5 years

Index

Authors' Vitae

Patrick T. Conley holds an A.B. from Providence College, an M.A. and Ph.D. from the University of Notre Dame, and a J.D. from Suffolk University Law School. His previous publications include *Democracy in Decline: Rhode Island's Constitutional Development, 1775-1841*; *Catholicism in Rhode Island: The Formative Era*; *An Album of Rhode Island History, 1636-1986*; and more than two dozen scholarly articles for history, law, and political science journals. A professor emeritus at Providence College, Dr. Conley now practices law and operates a real estate development business. From 1977 to 1984 he served as special assistant to Mayor Cianci, part of that time as chief of staff. In the early 1970s Dr. Conley was instrumental in the passage of a state statute establishing educational incentive pay for firemen and another lifting the local residency requirement.

Dr. Conley served as chairman of the Rhode Island Bicentennial Commission (ri76), chairman and founder of the Providence Heritage Commission, chairman and founder of the Rhode Island Publications Society, and general editor of the Rhode Island Ethnic Heritage Pamphlet Series. In 1977 he founded the Rhode Island Heritage Commission as a successor organization to ri76. Dr. Conley was also chairman of the Rhode Island Bicentennial [of the Constitution] Foundation and chairman of the U.S. Constitution Council. In May 1995 he was inducted into the Rhode Island Heritage Hall of Fame—one of a handful of living Rhode Islanders who have been accorded that honor.

Paul R. Campbell holds a B.A. and M.A. in history from Providence College and an M.L.S. from the University of Rhode Island. Mr. Campbell edited *Rights of Colonies Examined* by Stephen Hopkins and wrote, with Patrick Conley as coauthor, *Aurora Club: A Fifty Year History*. He has also published various articles relating to Rhode Island's history. Campbell was the library director of the Rhode Island Historical Society and coauthored with Dr. Conley the prize-winning *Providence: A Pictorial History*. He also directed the Providence Film Commission and plays a major role in directing the Providence Maritime Heritage Foundation, which operates the historic sloop *Providence*.

Mr. Campbell currently serves as the deputy director of policy for the City of Providence and participates on more than a dozen boards and commissions. He lives with his wife Susan on the East Side of Providence.